Understanding the Impact of Social Research on the Military

This book seeks to chart and evaluate the impact of social research on the military itself.

By "impact", the authors in this volume simply mean that which has a marked effect or influence on changing military policy, practices, knowledge, skills, behaviour, or living conditions. The book comprises a series of reflective contributions from scholars who have conducted research on the military as external scholars with no formal ties to the armed forces, as "native" researchers formally linked to them, as well as various kinds of contracted social scientists enabled by the military to carry out their investigations. The authors were asked to make the question of the impact of social scientific research on the armed forces an object of study in itself and to situate their reflections in terms of wider analytical questions. As a result, the chapters can be divided, broadly speaking, into two types of orientation: some are centered on theoretical and analytical issues, while others focus on the researchers' lived experiences.

This book will be of interest to students of military studies, sociology, organisational studies, psychology and political science.

Eyal Ben-Ari is a research fellow at the Institute for Society, Security and Peace, Kinneret College on the Sea of Galilee, Israel.

Helena Carreiras is director of the National Defense Institute, Portugal, and Associate Professor, University Institute of Lisbon.

Celso Castro is full professor and dean of the School of Social Sciences, Fundação Getulio Vargas, Brazil.

Cass Military Studies

Military Strategy of Middle Powers
Competing for Security, Influence and Status in the 21st Century
Håkan Edström and Jacob Westberg

Military Mission Formations and Hybrid Wars
New Sociological Perspectives
Thomas Vladimir Brond, Uzi Ben-Shalom, and Eyal Ben-Ari

French Defence Policy Since the End of the Cold War
Alice Pannier and Olivier Schmitt

War and International Relations
A Critical Analysis
Balazs Szanto

A Global History of Pre-Modern Warfare
Before the Rise of the West, 10,000 BCE–1500 CE
Kaushik Roy

Military Strategy of Great Powers
Managing Power Asymmetry and Structural Change in the 21st Century
Håkan Edström and Jacob Westberg

Counterinsurgency Warfare and Brutalisation
The Second Russian-Chechen War
Roberto Colombo and Emil Aslan Souleimanov

Managing Security
Concepts and Challenges
Laura R. Cleary and Roger Darby

Understanding the Impact of Social Research on the Military
Reflections and Critiques
Eyal Ben-Ari, Helena Carreiras, and Celso Castro

For more information about this series, please visit: https://www.routledge.com/Cass-Military-Studies/book-series/CMS

Understanding the Impact of Social Research on the Military

Reflections and Critiques

Edited by Eyal Ben-Ari, Helena Carreiras, and Celso Castro

Routledge
Taylor & Francis Group
LONDON AND NEW YORK

First published 2022
by Routledge
2 Park Square, Milton Park, Abingdon, Oxon OX14 4RN

and by Routledge
605 Third Avenue, New York, NY 10158

Routledge is an imprint of the Taylor & Francis Group, an informa business

British Library Cataloguing-in-Publication Data
A catalogue record for this book is available from the British Library

Library of Congress Cataloging-in-Publication Data
A catalog record has been requested for this book

ISBN: 978-0-367-76031-1 (hbk)
ISBN: 978-0-367-76039-7 (pbk)
ISBN: 978-1-003-16521-7 (ebk)

DOI: 10.4324/9781003165217

Typeset in Times New Roman
by MPS Limited, Dehradun

Contents

Figure

Tables

Acknowledgments

The organizers would like to thank the participants of the panel "The Impact of Sociological Research on the Military", carried out by RC01 at the XIX ISA World Congress of Sociology in Toronto, 2018, which gave rise to the idea of this book, and Verônica Azzi for the meticulous help in preparing the manuscript.

Contributors

Eyal Ben-Ari (PhD Cambridge University) was a professor of Anthropology at the Hebrew University of Jerusalem and is now a fellow of the Center for Society, Security and Peace at Kinneret College on the Sea of Galilee. He has carried out research in Israel, Japan, Singapore, Hong Kong and Taiwan. His main areas of research are the sociology of the armed forces, early childhood education, and popular culture in Asia. Among his recent books are *Japanese Encounters* (2018), (with Zev Lehrer, Uzi Ben-Shalom and Ariel Vainer) *Rethinking the Sociology of Warfare* (2010), (with Birgitte Refslund Sørensen) *Civil-Military Entanglements: Anthropological Perspectives* (2019), (with Nissim Otmazgin) *Creative Context* (2020), (with Thomas Brond and Uzi Ben-Shalom) *Military Mission Formations and Hybrid Wars* (2020). He has published articles spanning a variety of disciplines including the *American Ethnologist*, *Armed Forces and Society*, *Citizenship Studies*, *Cultural Anthropology*, *Religion*, *The Sociological Review*, and *Journal of Strategic Studies*.

Helena Carreiras is director of the Portuguese National Defense Institute and associate professor at ISCTE-Lisbon University Institute, where she served as dean of the School of Sociology and Public Policy. She holds a PhD in Social and Political Sciences from the European University Institute in Florence. In 2018–2019, she held the position of vice-president of the Institute for Social and Public Policies at ISCTE-IUL and served as president of ERGOMAS (European Research Group on Military and Society) for the period 2016–2019. Her research interests are in gender and society, armed forces, civil-military relations, security and defense and research methodology. Her work has focused on gender integration in military institutions and gender aspects of international security. She is the author of *Gender and the Military. Women in the Armed Forces of Western Democracies* (Routledge, 2006) and co-editor of *Researching the Military* (Routledge, 2016), *Qualitative Methods in Military Studies. Research Experiences and Challenges* (Routledge, 2013) and *Women in the Military and in Armed Conflict* (Vs Verlag, 2008).

Celso Castro is full professor and the dean of CPDOC, the School of Social Sciences at Fundação Getulio Vargas, Brazil. He holds an MA (1989) and a PhD (1995) in Social Anthropology from the Museu Nacional/UFRJ. Castro works on the intersection of Anthropology and History, especially regarding Brazil's history of the military. He has many publications on the topic, among which are *O espírito militar* (Zahar, 1990, 3rd ed. rev., 2021), *Os militares e a República* (Zahar, 1995), *A invenção do Exército brasileiro* (Zahar, 2002), *Antropologia dos militares* (FGV, 2009, co-org.), *Exército e nação* (FGV, 2012), *Qualitative Methods in Military Studies* (co-org, Routledge, 2013) and *Researching the Military* (co-org., Routledge, 2016). He was one of the editors of an eight-volume series about figures from the military in Brazilian post-1964 History. He also published and edited a series of classic anthropological and sociological texts in Portuguese and authored other volumes about the social scientific methods and theories. Castro is vice-president of the Research Committee on Armed Forces and Conflict Resolution of the International Sociological Association and extraordinary professor at Stellenbosch University, South Africa (2021–2023). He was a member of the Executive Board of the Brazilian Studies Association (BRASA, 2004–2008) and a visiting researcher or professor at the universities of Florida (1991–1992), Oxford (2000), Grenoble (2009) and El Colegio de México (2010).

Martin Elbe (Prof. Dr. Dipl.-Kfm. Dipl.-Soz.) is a senior researcher at the Bundeswehr Center for Military History and Social Sciences in Potsdam (Germany). He is chairman of the Working Group on the Military and Social Sciences in Germany. His work focuses on military sociology and social psychology (personnel, organization, health, and sports) and research methodology (empirical social research, organizational diagnosis, understanding). He has published more than 130 papers and 13 books. Most recent publications: Elbe, M. (2020, Ed.): *Die Gesundheit des Militärs.* Band 53 Militär und Sozialwissenschaften, hrsg. vom AMS. Baden-Baden: Nomos. [Military Health]; Elbe, M. & Erhard, U. (2020): *Konstruktive Organisationsentwicklung*: Menschen verstehen · Organisationen gestalten · Lernkulturen entwickeln. Baltmannsweiler: Schneider Verlag Hohengehren. [Constructive Organizational Development: Understanding People, Shaping Organizations, Developing Learning Cultures]; Elbe, M. (2019): Consequences of Leadership Behavior for Job Satisfaction and Health in the Bundeswehr. In: *Wehrmedizinische Monatsschrift.* 6(63), S. 186–193.; Elbe, M. (2018): Führung unter lernförderlichen Arbeitsbedingungen am Beispiel von Einsatzorganisationen. In: *Arbeit. Zeitschrift für Arbeitsforschung, Arbeitsgestaltung und Arbeitspolitik.* 4 (27), S. 345–367. [Leadership under Learning-Conducive Working Conditions on the Example of Emergency Organizations]

Gerhard Kuemmel (Dr. phil.) is a senior researcher at the Bundeswehr Center for Military History and Social Sciences in Potsdam (Germany). He studied Political Science, Sociology and History at the Philipps-University in Marburg, Germany, and received the J. William Fulbright Dissertation Award in 1996. From 2010 to 2014, he was the president of the Research

Committee 01: Armed Forces and Conflict Resolution (RC 01) of the International Sociological Association (ISA). He also serves on the Editorial Board of "Armed Forces & Society". He is Lecturer of Military Sociology in the Master Program "War and Conflict Studies" at Potsdam University. His research focuses on the integration of women into the armed forces, on military identity, on representations of the military in popular culture and, more recently, on veterans. His more recent publications in English include: "Debating German Special Forces: A Scandal in the Military, a Documentary, and a Thread". In: *Social Media and the Armed Forces.* Edited by Eva Moehlecke de Baseggio, Olivia Schneider & Tibor Szircsev Tresch, Cham: Springer Nature 2020, pp. 129–144; "Military Identity and Identity within the Military." In: *Handbook of the Sociology of the Military*, second revised edition. Edited by Giuseppe Caforio & Marina Nuciari, Cham: Springer Nature, 2018, pp. 477–493; and "The Bundeswehr and Female Soldiers: The Integration of Women into the Armed Forces (2000–2015)." In: *Connections: The Quarterly Journal*, Vol. 14, 2015, No. 3, pp. 61–86.

Piero Leirner is MsC (1995) and PhD (2001) in social anthropology from the University of São Paulo, Brazil. His research in this area was on military hierarchy, theory of war, ethnography of the military and on the relationship between social sciences, military and government agencies. He published, among others, *Meia-Volta, Volver: an anthropological study on the military hierarchy* (Ed. FGV, 1997) and with Celso Castro, *Anthropology of the Military: reflections on field research* (Ed. FGV, 2009). More recently, his research has turned to the topic of hybrid war, thinking especially of the "right turn" in Brazil. He published *Brazil on the Spectrum in the Hybrid War* (Ed. Alameda, 2020), and "Hybrid warfare in Brazil: The highest stage of the military insurgency" (*HAU: Journal of Ethnographyc Theory,* 10 (1), 2020).

Laura Masson is PhD (2007) and Master (1999) in Social Anthropology, Postgraduate Programme in Social Anthropology, Federal University of Rio de Janeiro (FURJ), Brazil. She is a professor at the National University of San Martín in Argentina. She was a member of the Gender Policy Council of the Ministry of Defence in Argentina from 2007 until 2019. She is the author of *Politic in female. Gender and power in Province of Buenos Aires* (2004), *Feminists everywhere. An ethnography of spaces and feminist narratives in Argentina* (2007) and editor of the book *Argentine Military Women. Evaluation of Gender Policies in the Defence Sphere* (2020). Coordinated the Dossier *Disruptive Violence. Women and the Use of Force* (2020), for the journal *Etnografías Contemporáneas.* She is currently director of Gender Policies at the Directorate of Human Rights and International Humanitarian Law of the Ministry of Defence of Argentine.

Soili Paananen is associate professor of Crisis Management at the Department of Leadership and Military Pedagogy, National Defence University. She holds a PhD degree in adult education. Her research interests and expertise include complexity in crisis management, military leadership, education and training for crisis management and peacekeeping operations, as well as sociomateriality in knowledge construction especially when situational awareness is developed. In recent years she has published in academic journals such as Management Learning and Scandinavian Journal of Military Studies.

David R. Segal earned his PhD in Sociology at the University of Chicago in 1967, spent 10 years on the faculty of the University of Michigan, and 40 years at the University of Maryland, where he founded and directed the Center for Research on Military Organization, and held courtesy appointments in the Department of Government and Politics and the School of Public Policy, and is now Professor Emeritus of Sociology and Distinguished Scholar-Teacher. He also had visiting appointments at the Army Research Institute, the Walter Reed Army Institute of Research, the Brookings Institution, West Point, the University of Bonn, and the University of Surrey. He has served as president of the Inter-University Seminar on Armed Forces & Society, the Research Committee on Armed Forces and Conflict Resolution of the International Sociological Association, and the District of Columbia Sociological Society, and as chair of the Section on Peace, War and Social Conflict of the American Sociological Association. He is the recipient of the Robin M. Williams Award from the American Sociological Association, the Robert M. Yerkes and Julius E. Uhlaner Awards from the American Psychological Association, The Morris Rosenberg and Stewart A. Rice Awards from the District of Columbia Sociological Society, the Morris Janowitz Award from the Inter-University Seminar on Armed Forces & Society, the Distinguished Former Faculty Award from West Point, and the Alumni Distinguished Career Award from the University of Chicago. He has also received the American Sociological Association Award for Public Understanding of Sociology, and has twice been awarded the U.S. Army Medal for Outstanding Civilian Service. His books include *Military Sociology* (4 vols. With James Burk), *The Postmodern Military* (with Charles C. Moskos & John A. Williams), *Peacekeepers and their Wives* (with Mady W. Segal), *Recruiting for Uncle Sam*, and *The All-Volunteer Force* (with Jerald G. Bachman and John D. Blair).

Introduction: social science, influence, and change in the military

Eyal Ben-Ari, Helena Carreiras, and Celso Castro

Despite a significant growth of studies on methods for studying the armed forces, sustained studies on the impact of the actual effects of research on the military are still all too rare. Indeed, the past decade has seen the publication of no fewer than four hefty edited collections (Carreiras et al., 2016; Castro & Carreiras, 2013; Williams et al., 2016) that explore and explain the various methods, research tools, interpretive techniques, and modes of theory building involved in carrying out research on the military. Against the background of the long historical interest of sociologists in the armed forces (going back to Max Weber and continuing during WWII), these four volumes reflect the emergence of a relatively new reflexive mood in the field, one that entails a pause to think and rethink the manner by which scholars can approach the study of the armed forces in the most fruitful manner.

Our volume, however, proposes an unusual and hopefully innovative perspective that seeks to fill a gap in studies of (and for) military forces by charting out and evaluating the impact of social research on the military itself. By impact, we simply mean having a marked effect or influence on changing military policy, practices, knowledge, skills, behavior, or living conditions. As such, our volume complements rather than competes with previous volumes on methods. Concretely, this edited volume comprises a series of reflective contributions from scholars who have conducted research on the military as external scholars with no formal ties to the armed forces, as 'native' researchers formally linked to them, as well as various kinds of contracted social scientists enabled by the military to carry out their investigations.

To be sure, studying the impact of social scientific research on the armed forces is similar to studying the influence of such research on any large, public organization. But the military is both like and unlike other such organizations. It is similar to other large public bureaucracies – such as the educational system, welfare structures, or hospital administrations – in that organizational change involves untidy processes in which multiple actors with multiple agendas (including those of the researchers themselves) participate, often rendering it difficult to measure or appraise the impact of specific research projects on policy or practices. And indeed, as the chapters

DOI: 10.4324/9781003165217-101

in this volume underscore (see Kummel & Elbe; Masson, Carreiras, Paananen, this volume), the impact of sociological research on the military is refracted through complex processes of production, dissemination, and consumption of knowledge via often opaque and complex dynamics of acceptance (or partial rebuff or outright rejection), understanding (often partial), interpretation (according to different agendas), and only then (at times), actual implementation based on social scientific findings.

And yet the armed forces are unlike other large public organizations because their uniqueness lies in handling legitimate (if sometimes contested) state violence (Boene 1990). Specializing in organized state-mandated violence entails a host of characteristics that arguably make an appraisal of social science's influence on the military even more opaque than assessing its influence on other large organizations. This difficulty arises because of the secrecy and control of information about changes involved in "national security", the loyalty and commitment of military members, the suspicion (Caforio & Nuciari, 2006) or distrust (Kummel & Elbe) toward outsiders, the perceived public sensitivity toward the use of armed force, and the possible seduction of researchers by senior generals (Robben, 1995). In addition to issues of sensitivity, the configuration linking principles such as 'hierarchy' and 'discipline' in the armed forces create difficulties that are different from those found in other institutions. All these factors make it both difficult to ascertain the impact of research and may at times create barriers to actually effecting change.

Before going on to set out the main clusters of issues that the individual chapters and the volume as a whole deal with, let us clearly spell out the parameters of our volume. We use sociology or social science as a shorthand for the study of the military by members of sociology and its closely allied disciplines of anthropology, social psychology, or parts of political science and geography. Accordingly, we have left out of our purview two related kinds of research. We did not include macro-studies about the impact of political scientists or experts in international relations on national strategy, security doctrine, or the management of international conflicts (see Desch, 2019; Levy, 2020; Morrissey, 2011). Rather, we include studies investigating the influence on military organization, social processes-centered combat units, and civil-military relations. Conversely, at the micro-level, we do not cover examinations of the effects of the therapeutic parts of psychology – allied as they are with the mental health disciplines (Ben-Shalom & Fox 2009) – since they follow a rather different trajectory of impact than the more sociological studies (including, again, social or organizational psychologies). Similarly, we have not included work on social scientists working *with* the military – as in the historical use of social sciences by the US security forces (Price, 2008, 2011a, 2011b; Tomforde, 2011) and the newer and controversial American Human Terrain System groups (McFate, 2006, 2019; McFate & Laurence, 2015). Finally, and to further clarify, we do not refer to the many studies of the military that simply use the access and availability of soldiers to explore various theoretical questions without an orientation toward change in the armed forces (for example, Goodwin et al., 2018). The reason for this exclusion is the fact that these kinds of studies are similar to the way psychologists utilize the availability of university students to explore theoretical problems.

To be candid, in many cases we simply do not know how, when, or whether at all, any social scientific knowledge is used by parts of militaries. Yet this realization should not deter us from attempting to set out the issues involved in effecting change. Hence, all of the essays are reflective in one way or another. Since most researchers do not document the routes by which their work has had an effect on the armed forces, we have asked all of the contributors to "take a step back" from their own ongoing work and to think about these issues as they emerged throughout their careers or the histories of their institutions and disciplines. In other words, we asked the authors to make the question of the impact of social scientific research on the armed forces an object of study in itself and to situate their reflections in terms of wider analytical questions. As a result, the chapters can be divided, broadly speaking, between two kinds of complementary orientations: some are centered on more theoretical and analytical issues, while others focus on the researchers' lived experiences.

Three types of knowledge

Analyzing the impact of social research on the armed forces involves situating scholars within the social dynamics of an often-bewildering array of organizational, institutional, and social networks and relationships centered on the military. Furthermore, this essential social character of influence prods us to explore not only direct influence on policies and practices but how our very participation as researchers and social science experts (of all kinds) with the military carries in and of itself a host of consequences that involve organizational change. We propose that using the typology of three types of *knowledge-constitutive interests* developed by Habermas' (1971; Ben-Ari & Levy, 2014) offers a fruitful entryway into the issues dealt with by this volume. It is such an entryway because it suggests how each of these ideal types is expressed in a particular type of scholarly inquiry and orientation toward social action and thus can illuminate the diverse ways in which social research shapes the armed forces (and is shaped) by its influence on them. Let us set out the ideal type of each interest, exemplify it in regard to the study of the military, and suggest its relation to different types of influence and change.

The engineering model

The first interest suggested by Habermas is a *technical interest* in the prediction and control of the natural environment and society that aims at tying knowledge-production to controlled observation and testable general explanations, yielding the empirical-analytic sciences. We find that this kind of knowledge-constitutive knowledge undergirds what Moskos (1988) called the engineering model of military sociology, a knowledge oriented to seeking explanatory regularities by which military commanders can effectively control the actions of the armed forces or parts of them. The idea – which is probably still the most dominant one in research about the armed forces – is reaching optimum levels of functioning of the military as a whole or parts of it (see Ben-Ari, Kummel & Elbe; Paananen; Segal, this volume). That this is an engineering model is not surprising since much of post-World War II sociology borrowed its governing theoretical armature from industry and emphasized the certainty created by control and the manipulation of variables to achieve predictability and surety (Ben-Ari et al., 2010; Brond et al. 2020). This older sociology of tactical units developed during World War II was honed during the subsequent decades and was led by a union between psychology (especially social-psychology) and the military. The key questions in the classic sociology of combat and combat units are centered on the contribution of different elements of social structure to the survivability and smooth operation of military units – basically resilience and efficacy.

Today, research that is turned into applied knowledge is guided by such technical interests, and this kind of work typifies many of the studies carried out by scholars in military academies and the research units of militaries and defense ministries. These actors provide resources for decision-making regarding such issues as cohesion and morale (Balkind, 1978), and soldierly, veterans' and families well-being. During the past two decades, the influence of positive psychology has been felt in studies of resilience and hardiness (Matthews, 2014). Across the board, commanders see such knowledge as "useful" and at times take it into account in guiding policy and practice. In addition, in the US scholars have even provided "beneficial" advice about how to frame the use of force to muster support despite mounting casualties (Gelpi et al., 2009). More widely, militaries have used this kind of functional knowledge in regard to Human Resources Management (recruitment and retention, gender integration and the management of diversity, and training), organizational and management issues (work division, job design, organizational culture), and legitimacy and reputation issues (social support among wider populations). At the same time, however, the armed forces have very rarely commissioned social research about planning, guiding, and evaluating military operations.

The enlightenment model

The second *practical human interest* in establishing consensus makes use of the cultural-hermeneutic approach relying on interpretive methods.

Assuming that social life is constituted by social actions and that actions are meaningful to the actors and to the other social participants, the scholarly task is to interpret the meanings of social actions. It aims at attaining reliable intersubjective understanding established in the practice of ordinary language to assure channels of communication, by expanding possibilities of mutual and self-understanding in the conduct of life. The search for this kind of knowledge is what Moskos (1988) calls the enlightenment model of the sociology of the military.

While this kind of interest is not directly aimed at influencing the armed forces, we propose that it nevertheless underlies two kinds of (often potential) impact of social scientists on the armed forces. The first characterizes at least some efforts undertaken by social scientists in military educational institutions either when they are hired as faculty (Ender & Hajjar, 2019; Fujimura, 2003) or when external academics give lectures about various issues. The work they carry out involves educating cadets or soldiers and about key social scientific fields (psychology, political science, management, sociology, or anthropology), illuminating social issues, or providing scholarly perspectives on civil-military relations. This kind of educational effort is squarely embedded in the enlightenment ideas of broadening the minds of soldiers and commanders with understandings about individuals, societies, and cultures. Students are beneficiaries in the sense of internalizing (to different degrees) ideas about how social scientists approach social issues in their analyses. In a closely related manner, the participation of social scientists in decision-making arenas is often aimed at informing and enlightening policy-makers (Ben-Ari, Kummel & Elbe; Masson, Paananen; Segal, in the volume).

The second kind of enlightenment-related influence that has developed to a great extent for the past two decades is teaching cultural awareness and intercultural competencies inside and outside military schools. While involving political and ethical problems, today's militaries see a need to prepare soldiers for missions abroad in conflicts typically based on complex ethnic, religious, economic, and political dynamics. In anglophone countries, cultural intelligence training is aimed at language and cognitive skills apt for various sociocultural settings. However, the American term "cultural intelligence" is problematic as it suggests that it is part of the intelligence field and not everyone's responsibility by way of acting sensibly in other cultures (McFate, 2005; Holmes-Eber et al., 2009; Cline, 2017; Parenteau, 2020). In European contexts, scholars usually refer to cultural awareness and intercultural competencies that military leaders of all levels need in order to succeed in socioculturally challenging situations (Haugegaard, 2020; Masson & Moelker, 2020; Tomforde, 2020). The fact that these efforts are part of an enlightenment orientation is best encapsulated in Cline's (2017) argument for a move from cultural intelligence to cultural understanding.

The critical sociological model

In contrast to the former two sets of interests, the goal of the *third* category, the *emancipatory interest,* is to achieve emancipatory knowledge by counteracting the oppressive effects of the social construction of knowledge. Yagil Levy (2015, 2020) has formulated this orientation most clearly in regard to the armed forces in regard to critical military sociology. Oriented toward the general public, this kind of knowledge-constituting interest is directed toward initiating debates to raise awareness of military-related issues (such as the management of diversity, gender-related issues, the effects of personnel policies or civilian control) so as to monitor the ways by which states manage organized violence and use the military for other goals. Critical thinking identifies constraining structures of power, such as relations of dependence, that unreflectively appear as natural.

Many of the social scientific studies of the military bearing this kind of critique began to be produced from about the mid-1960s in the wake of the Vietnam War and the associated peace and student movements of the time. Led by Tilly (1992) or Giddens (1985), scholars argued that war and the institutions of war making are integral to the creation of states and to the mobilization of social resources. Such scholars have done much to uncover the main social and, especially, political mechanisms by which war has become part and parcel of the very dynamics of contemporary countries. Accordingly, critical thinking in the domain of civil-military relations has traditionally focused mainly on the rise of tacit forms of militarism and their contribution to the construction of power relations (for instance, Levy, 2015; Lutz, 2001, 2002). Another veritable outpouring of scholarly literature – often lead by work published in the journal *Critical Military Studies* – involves how the armed forces entrench ethnic, race, or (most important for feminist scholars) gender relations (Higate, 2003; Basham & Bulmer, 2017; Woodward & Duncanson, 2017). However, much of this scholarly literature, it should be clarified, is aimed at the social science disciplines themselves rather than at the armed forces. At the same time, and while a relatively recent addition, for the past decade or so critical geography (Rech et al., 2015) has emerged as a field of its own analyzing such issues as the siting of camps and bases or the role of military geography in nation and state-building (Barton & Irarrázaval, 2014; Basham 2013; Gillem 2007; Woodward 2005),

While the beneficiaries of such studies are of course academics, it can be argued that some such studies have had a direct effect on the armed forces of the industrial democracies. One example is Winslow's (1997) work on the Canadian Airborne Regiment in Somalia that led to a wholesale reform of this unit and of the educational curriculum of the Canadian forces more generally. Similarly, the initial struggles in and around gender integration are a good example of how critical approaches outside the military have led to changed policies (a theme we will return to later).

Often, however, scholars producing emancipatory knowledge encounter resistance from military commanders, especially top commanders who may see such knowledge as threatening the very legitimacy of the armed forces and their "good image". Indeed, underpinning such perceptions may be the military cosmology that divides the world between "friends" and "enemies". In addition, in many countries and over different periods of time, the military has participated (or is still participating) actively in politics, thus adding new (and sometimes dramatic) dimensions to the so-called "civil-military relations".

In some cases, we propose that a greater understanding of critical work on the military may actually facilitate a form of cooptation of ideas by commanders who understand the power of such critiques.

However, for all this, we follow Calhoun's (2010) observations that different disciplines have different attitudes in terms of working with the military since, in our terms, they produce different kinds of knowledge. For instance, on the whole, psychologists tend to be less, and many sociologists tend to be more, critical of the armed forces. And, in a related manner, national differences in histories shape the kinds of effects that social science may have on the armed forces.

Three distinctions

In this section, we extend our analysis by linking it to three somewhat overlapping distinctions that we believe can illuminate how the three orientations toward knowledge are related to how, and whether, social research impacts the armed forces.

Explicit and implicit goals

The first distinction we offer is between explicit (sometimes formally designated) and implicit (at times conceptualized as informal) aims of participants. Among the formal aims of research would be goals such as analyzing the effects of changed policy or documenting and analyzing changing motivations to enlist or remain in service. These formal, explicit goals are usually designated in legal contracts or documents. The informal aims may entail such goals as furthering academic knowledge, promoting scholarly careers, or analyzing issues not formally dictated by the military institution (Ben-Ari; Castro; Leirner, this volume). These implicit goals may *or* may not be stated and may be either part of the informal set of agreements between the participants or hidden from the view of one of the partners.

It is here that the diverging or complementary agendas of military personnel and researchers (external and internal, contracted or enabled) may be in tension and could shape the impact of research or how this impact is negotiated. Scholars (Jans, 2014; Mosser, 2010; Segal, this volume) have suggested that the problems of contact between these two groups are essentially different communication styles or ways of formulating problems and action. By looking at implicit goals, however, it becomes clear that these are also distinguished by agendas that are taken for granted for them. Researchers (including ourselves) are not just producers and mediators of knowledge between the military and civilian society, but rather become an extension of the two domains, serving *multiple* agendas and aims – at times at one and the same time (Kummel and Elbe; Paananen, this volume).

Direct and indirect influence

The second distinction is between direct versus indirect use of the sociological knowledge produced. The direct effects of research, which are very much part of the engineering model of social science, are ostensibly easy to detect through such indicators as changed policies or organizational structures and procedures (for instance, the establishment of a new research unit or changed criteria for admittance to military roles). While it may not be easy to trace the trajectory of influence, this kind of impact has been the most commented upon pointed out by scholars, and the one we usually look for to talk about the influence of producing knowledge about the armed forces.

But there may also be indirect impacts on diverse actors. One example (to which we shall return) may be new research-driven conceptualizations or theoretical frameworks that enrich understandings of the armed forces. Another indirect influence, one related to the enlightenment and critical sociological models, may be a raised awareness of social issues emerging out of investigations and deemed politically important for activists such as human rights lawyers or feminists (and some scholars). These indirect uses and impacts, in turn, may again be the source of tensions between scholars and military commanders. In a closely related way, tensions inside the academic community of social scientists are often expressed in the exchange of accusations that are not strictly intellectual, such as being 'co-opted' or being 'anti-military'.

Intended and unintended consequences

The third distinction we introduce, and perhaps the most interesting, is the one between intended and unintended impacts. Primary among the intended consequences are improvements brought about by the engineering model (see some of the examples in Paananen; Kummel & Elbe; Ben-Ari; Segal, this volume). The intended consequences of the enlightenment model are more difficult to appraise but can be obtained via the self-reflection of commanders and soldiers. It is, however, in the unintended consequences that wider effects of sociological research can be seen. Let us provide a number of examples of such unintended (or, alternatively, unplanned, unexpected, or unforeseen) consequences. The first is centered on the implications of the engineering model. If social scientists are concerned about not causing harm or limiting it (Calhoun 2010), then a consequence of their successful actions may be the improvement of the functioning of the armed forces so that they "deliver" violence in an ever more effective manner that is actually unforeseen by scholars.

The second unintended consequence is the very participation of scholars and the presentation of research, which may act to ornament a certain meeting. In some cases, intellectuals and professors may be used simply as a background for politicians or senior civil servants as they award them with an academic aura. Thus, while scholars doing research, presenting their findings of providing counsel to senior decision-makers may intend certain outcomes, the unintended consequence of their very participation in and of itself (regardless of the knowledge they are trying to impart) grants a meeting or a decision an academic mark. The contribution by Ben-Ari illuminates this point through the account of how he was invited by the Israeli military to participate in meeting simply as a "professor", without mention of his expertise.

The third of these consequences is the fact that the participation of scholars serves to legitimate a certain policy or practical change. In their chapter, Kummel and Elbe relate the case of how the German Federal Minister of Defense cited social research to make its point. What they underscore further is that this instance is but part of a more general process of the scientification of the social. That way, policy makers make use of social scientific theories and findings in ways that support their predetermined – or at least preconceived – views by borrowing their authority as professional experts and of scientific knowledge.

The fourth consequence, and in a different manner in Leirner and Castro's chapters, the very acts of having social scientists give lectures, attend meetings, and write books is but part of the unintended legitimation of certain military units (and therefore, perhaps, their actions) but not of policy. Their accounts highlight a much wider potential resonance of academic participation. In Castro's report, the book he wrote about the Brazilian military academy took on a life (or career) of its own. The volume became a totem for the identity and memory of not only the class of cadets that he studied, but more generally, of the academy. Furthermore, the book (again, without necessarily making any reference to its findings and conclusions) could then be vastly bought, read, cited, exchanged, or displayed as a token (perhaps a mascot) of this unit. For Leirner, the lecture he gave to members of the military police of the state of São Paulo (Brazil) was similarly turned into a ritual of affirmation of the unit and its members.

A fifth consequence (that may be partially intended) involves raising the level of in-house research by setting academic standards. Accordingly, through such mechanisms, as the integration of civilian academic researchers into military institutions or awarding dual positions to researchers in military and civilian institutions of higher education (Ben-Ari, 2016, in this volume), scholars have raised these academic standards? have a host of consequences. These include bringing in new formulations, concepts or approaches, or subjecting internal research to external criteria (for all of its faults peer review is still a sort of academic equivalent of quality control). Paananen's chapter exemplified this point by showing how in-house research and teaching is subject to review by the Finnish Educational Evaluation Centre.

The sixth and final consequence which we find intriguing (and ironic) is one unintended consequence of the enlightenment model for officers, namely that today many of them have internalized – via courses in management and business schools, popular culture, the work of consultants, or military educational establishments – the language of some of the social sciences. On the one hand, this development is intentional and a positive outcome of the educational process. On the other hand, because they can "talk the talk", that is, speak fluently or convincingly about something or in a way intended to please or impress others, they can more easily coopt social scientific research and social scientists and depoliticize issues. They may, then, curtail the already problematic independence of in-house scholars (Kummel & Elbe; Paananen, Masson, this volume). Moreover, and again paradoxically, the very scholars who object to and criticize the military may have their work used in the educational frameworks of the armed forces.

In lieu of a conclusion: a "Messy" model of influence

We now turn to a final proposal that is based on our impressions and our (by now) collectively tens of decades of experiences of working with or on the study of the military with scholars, military personnel, members of NGOs, representatives of the media, and politicians. Simply put, we suggest that the relations between social science and policy-making, practice, and implementation can be understood via a coalitional or ("messy") interactive-network model that is marked by multiple trajectories of entanglements between civilian and military spheres (Sørensen & Ben-Ari, 2019).

The picture of policy and implementation that emerges from our impressions and the chapters to this volume is often of a disorderly assortment of interconnections that mix competition and cooperation, frequently marked by a constant "back-and-forth-ness" between decisions and implementation, and characterized by unforeseen results. In this depiction military decision-makers (like all such decision-makers) are nodes around which swirl a host of actors, such as fellow soldiers (superordinates, peers, and subordinates), friends and allies (and enemies or competitors), political overlords, and (sometimes) consultants, representatives of the media, and (in-house and external) social scientists. Each of these groups is characterized by its own world-view, assumptions about what is worthy of action and organizational practices. Further still, each group is far from homogenous and, for example, within the armed forces, there are tensions between services, arms, and units and among academics; conflict is very much part of the way knowledge is produced. The processual and essential negotiated character of influence is brought out most clearly in the chapters by Paananen, Kummel and Elbe, Masson, and Carreiras. Sometimes, it is (often rather loose) coalitions that scholars belong to that bring about change. In such alliances, each actor provides different resources (such as legitimacy, authority, publicity, material means) that cumulatively bring about some kind of impact. Moreover, and especially since the armed forces are not homogenous, social scientists may provide support and power for opposition to military policy by allying with actors within the armed forces, some of whom become patrons. For military leaders, moreover, the reception of social scientific analyses is itself a set of processes that involve various forms of acceptance, interpretation, adaptation, or, in fact, disregard, or even rejection.

One of the clearest examples of the impact of social research on the armed forces has been the cluster of issues related to gender integration, as described in the chapters by Carreiras and Masson. The processes by which gender integration (as partial as it may be) is the outcome of untidy processes centered on – sometimes loosely and sometimes tightly coupled – cooperation among some senior commanders, academics, journalists, feminist lobbies, and politicians. Academics and their work were hence part of a wider movement that led to influence on the armed forces and eventually to some changes in regard to gender integration. These changes also exemplify the effects of critical sociology in the sense that Levy (2015) proposes: academics' engagement with wide publics (again along with other actors) seems to have been part of the process leading to transformed policies. What the contributions by Carreiras and Masson make clear, however, is that social scientists, even critical ones, not only need to direct their attention to informing publics but may, in many contexts, also benefit from direct links with the armed forces.

Conclusion

By way of conclusion let us make one point. While we as social scientists undertake many roles in our relations with the armed forces, one of the most important ones is that of scholarly experts. As such experts, we have the power to frame ideas about what is "worthy" of study and via that what is worthy of being analyzed, promoted, denigrated, or even simply read. As the contributions to this volume all underscore, when working with the military this power is always negotiated within wider organizational and social fields. Hence the impact of our work – its influence on policies, institutional arrangements, or the actions of soldiers – involves a plethora of effects that may be variously intended or unintended, and explicit or implicit but is always political,

References

Balkind, J. (1978). A Critique of Military Sociology: Lessons for Vietnam. *Journal of Strategic Studies* 1: 237–259.

Barton, J., & F. Irarrázaval (2014). Geographical Representations: The Role of the Military in the Development of Contemporary Chilean Geography. *Area* 46 (2): 129–136.

Bartone, P. T. (2006). Resilience Under Military Operational Stress: Can Leaders Influence Hardiness? *Military Psychology* 18 (Suppl.): 131–148.

Basham, V. (2013). *War, Identity and the Liberal State: Everyday Experiences of the Geopolitical in the Armed Forces*. London: Routledge.

Basham, V. M., & Bulmer, S. (2017). Critical Military Studies as Method: An Approach to Studying Gender and the Military. In R. Woodward, & C. Duncanson (Eds.), *The Palgrave International Handbook of Gender and the Military*. London: Palgrave Macmillan.

Ben-Ari, E. (1998). *Mastering Soldiers*. New York: Berghahn Books.

Ben-Ari, E. (2011). Anthropological Research, and State Violence: Some Observations of an Israeli Anthropologist. In L. A. McNamara, & R. A. Rubinstein (Eds.), *Dangerous Liaisons: Anthropologists and the National Security State* (pp. 167–184). Santa Fe: School of Advanced Research Press.

Ben-Ari, E. (2016). What is Worthy of Study about the Military? The Sociology of Militaries-in-Use in Current-Day Conflicts. In H. Carreiras, C. Castro, & S. Frederic (Eds.), *Researching the Military*. London: Routledge.

Ben-Ari, E., Lehrer, Z., Ben-Shalom, U., & Vainer, A. (2010). *Rethinking Contemporary Warfare: A Sociological View of the Al-Aqsa Intifada*. Albany: State University of New York Press.

Ben-Ari, E., & Levy, Y. (2014). Getting Access to the Field: Insider/Outsider Perspectives In J. Soeters, P. Shields, & S. Rietjens. (Eds.), *Routledge Handbook of Methods in Military Research* (pp. 9–18). London: Routledge.

Ben-Ari, E., & Lomsky-Feder, E. (1999). Introductory Essay: Cultural Constructions of War and the Military in Israel. In E. Lomsky-Feder, & E. Ben-Ari (Eds.), *The Military and Militarism in Israeli Society* (pp. 1–34). Albany: State University of New York Press.

Ben-Shalom, U., & Fox, S. (2009). Military Psychology in the Israeli Defense Forces: A Perspective of Continuity and Change. *Armed Forces and Society* 36 (1): 103–119.

Boene, B. (1990). How Unique Should the Military Be? A Review of Representative Literature. *Archive Eurepeennes de Sociologie* 31 (1): 3–59.

Boene, B. (2000). Social Science Research, War and the Military in the United States: An Outsider's View of the Field's Dominant Tradition. In G. Kummel, & A. D. Prufert (Eds.), M*ilitary Sociology: The Richness of a Discipline* (pp. 149–253). Baden-Baden: Nomos.

Brond, T., Ben-Shalom, U., & Ben-Ari, E. (2020). *Military Mission Formations and Hybrid Wars*. London: Routledge.

Caforio, G. (2007). Introduction: The Interdisciplinary and Cross-national Character of Social Studies on the Military – The Need for Such an Approach. In G. Caforio (Ed.), *Social Sciences and the Military: An Interdisciplinary Overview* (pp. 1–20).London: Routledge.

Caforio, G., & Nuciari, M. (2006). Social Research and the Military: A Cross-national Expert Survey. In Caforio, G. (Ed.), *Handbook of Sociology of the Military* (pp. 27–59). New York: Springer.

Calhoun, C. (2010). Social Science Research and Military Agendas: Safe Distance and Bridging a Troubling Divide. *Perspectives on Politics* 8 (4): 1101–1106.

Capshew, J. H. (1999). *Psychologists on the March: Science, Practice and Professional Identity in America, 1929–1969*. Cambridge: Cambridge University Press.

Carreiras, H., Castro, C., & Frederic, S. (Eds.) (2016). *Researching the Military*. London; New York: Routledge.

Castro, C., & Carreiras, H. (Eds.) (2013). *Qualitative Methods in Military Studies*. Oxford: Routledge.

Cline, L. (2017). From Cultural Intelligence to Cultural Understanding: A Modest Proposal. *Small Wars Journal*, 04/20/2017, https://smallwarsjournal.com/jrnl/art/from-cultural-intelligence-to-cultural-understanding-a-modest-proposal (last accessed 29 June 2020).

Denskus, T. (2016). From Social Movement to Ritualized Conference Spaces: Peace Research and Knowledge Production in Germany. *Peace & Change* 41 (3): 302–328.

Desch, M. (2019). *Cult of the Irrelevant: The Waning Influence of Social Science on National Security*. Princeton: Princeton University Press.

Ender, G. M., & Hajjar, R. M. (2019). Teaching War from a Sociological Perspective: The West Point Exemplar. *Res Militaris* vol. 9, n°1, Winter-Spring. http://resmilitaris.net/index.php?ID=1029407. Retrieved 14 January 2020.

Fosher, K. (2013). Reflections on Current Research: Science and Scientists in Military Organizations. *Journal of Culture, Language, and International Security* 1 (2). Accessed 24 June 2016.

Fujimura, C. (2003). Integrating Diversity and Understanding the Other at the U.S. Naval Academy. In P. R. Frese, & M. C. Harrell (Eds.), *Anthropology and the United States Military* (pp. 135–146). New York: Palgrave Macmillan.

Gelpi, C., Feaver, P. D., & Reifler, J. 2009. *Paying the Human Costs of War: American Public Opinion and Casualties in Military Conflicts*. Princeton, NJ: Princeton University Press.

Giddens, A. (1985). *The Nation-State and Violence*. Cambridge: Polity.

Gillem, M. (2007). *America Town: Building the Outposts of Empire*. Minneapolis: University of Minnesota Press.

Goodwin, G. F., Blacksmith, N., & Coats, M.R. (2018). The Science of Teams in the Military: Contributions from Over 60 Years of Research. *American Psychologist*, 73 (4):322–333.

Habermas, J. (1971). *Knowledge and Human Interests*. Boston: Beacon.

Haugegaard, R. (2020). Culture as Operational Enabler: Training Danish Officers to understand the Interaction between cultural Dynamics and Military Operations. In K. Enstad, & P. Holmes-Eber (Eds.), *Warriors or Peacekeepers? Building Military Cultural Competence* (pp. 97–116). New York: Springer.

Higate, P. R. (2003). *Military Masculinities: Identity and the State*. Westport: Praeger

Holmes-Eber, P. (2014). *Culture in Conflict: Irregular Warfare, Culture Policy, and the Marine Corps*. Stanford: Stanford University Press.

Holmes-Eber, P. et al. (Eds.) (2009). *Applications in Operational Cultural: Perspectives from the Field*. Quantico, Virginia: Marine Corps University Press.

Jans, N. (2014). Getting on the Same Net: How the Theory-driven Academic Can Better Communicate with the Pragmatic Military Client. In J. Soeters, P. M. Shields, & S. Rietjens (Eds.), *Routledge Handbook of Research Methods in Military Studies* (pp. 19–28). London: Routledge.

Levy, Y. (2015). Time for a Critical Military Sociology. *Res Militaris* 5 (2). http://resmilitaris.net/index.php?ID=1022274. Accessed 15 July 2020.

Levy, Y. (2020). What is the Social Responsibility of Social Scientists to Influence National Security Affairs? *Armed Forces and Society* (forthcoming).

Lutz, C. (2001). *Homefront: The Military City and the American Twentieth Century*. Boston: Beacon Press.

Lutz, C. (2002). Making War at Home in the United States: Militarization and the Current Crisis. *American Anthropologist* 104 (3): 723–735.

Masson, L., & Moelker, R. (2020). Teaching Gender, Teaching Culture: A Comparative Study of Gendered Dilemmas in Culturally Complex Situations. In K. Enstad, & P. Holmes-Eber (Eds.), *Warriors or Peacekeepers? Building Military Cultural Competence* (pp. 161–182). New York: Springer.

Matthews, M. D. (2014). *Head Strong: How Psychology Is Revolutionizing War*. Oxford: Oxford University Press.

McFate, M. (2005). Anthropology and Counterinsurgency: The Strange Story of Their Curious Relationship. *Military Review* 85 (2): 24–38.

McFate, M. (2019). *Military Anthropology: Soldiers, Scholars and Subjects at the Margins of Empire*. Oxford: Oxford University Press.

McFate, M., & Laurence, J. H. (Eds.) (2015). *Social Science Goes to War*. Oxford: Oxford University Press.

Morrissey, J. (2011). Architects of Empire: The Military-strategic Studies Complex and the Scripting of US National Security. *Antipodes* 43 (2): 435–470.

Moskos, C. (1971). *The American Enlisted Man: The Rank and File in Today's Military*. New York: Russell Sage Foundation.

Moskos, C. (1988) *Soldiers and Sociology. SLA Marshall Lecture*. Washington, D.C.: US Army Research Institute for the Behavioral and Social Sciences.

Mosser, M. W. (2010). Puzzles versus Problems: The Alleged Disconnect between Academics and Military Practitioners. *Perspectives on Politics* 8 (4): 1077–1086.

Parenteau, D. (2020). Cultural Inteliigence as Part of Officer's Virtue. In K. Enstad, & P. Holmes-Eber (Eds.), *Warriors or Peacekeepers? Building Military Cultural Competence* (pp. 11–22). New York: Springer.

Price, D. (2008). *Anthropological Intelligence*. Durham, NC: Duke University Press.

Price, D. (2011a). Counterinsurgency, Vietnam, Thailand and the Political Use of Militarized Anthropology. In L. McNamara, & R. Rubenstein (Eds.), *Dangerous Liaisons: Anthropologists and the National Security State* (pp. 51–76). Santa Fe: SAR Press.

Price, D. (2011b). *Weaponizing Anthropology*. Petrolia, CA: AK Press.

Rech, M., Bos D., Jenkings K. N., Williams, A., & Woodward R. (2015). Geography, Military Geography and Critical Military Studies. *Critical Military Studies* 1 (1):47–60.

Robben, A. (1995). The Politics of Truth and Emotion among Victims and Perpetrators of Violence. In C. Nordstrom, & A. C. G. M. Robben (Eds.), *Fieldwork under Fire* (pp. 81–103). Berkeley: University of California Press.

Schwartz, T. P., & Marsh, R. M. (1999). The American Soldier Studies of WWII: A 50th Anniversary Commemorative. *Journal of Political and Military Sociology*, 27 (1): 21–37.

Segal, D. R. (2007) Current Developments and Trends in Social Research on the Military. In G. Caforio (Ed.), *Social Science and the Military* (pp. 46–65). London: Routledge.

Soeters, J., Shields, P. M., & Rietjens, S. (Eds.) (2014). *Routledge Handbook of Research Methods in Military Studies*. London: Routledge.

Sørensen, B. R., & Ben-Ari, E. (2019). Rethinking Civil-Military Connections: From Relations to Entanglements. In B. R. Sørensen, & E. Ben-Ari (Eds.), *Civil-Military Entanglements* (pp. 1–20). Oxford: Berghahn.

Tilly, C. (1992). *Coercion, Capital, and European States, AD 990–1992*. Cambridge, MA: Basil Blackwell.

Tomforde, M. (2011). Should Anthropologists Provide Their Knowledge to the Military? An Ethical Discourse taking Germany as an Example. In L. McNamara, & R. Rubinstein (Eds.), *Dangerous Liaisons: Anthropologists and the National Security State* (pp. 77–100). Santa Fé: SAR Press.

Tomforde, M. (2020). Intercultural Competencies in the Bundeswehr: Officer Training and Mission Realities. In K. Enstad, & P. Holmes-Eber (Eds.), *Warriors or Peacekeepers? Building Military Cultural Competence* (pp. 139–160). New York: Springer.

Williams, A. J., Jenkings, K. N., Woodward, R., & Rech, M. F. (Eds.) (2016). *The Routledge Companion to Military Research Methods*. London: Routledge.

Winslow, D. (1997). *The Canadian Airborne Regiment in Somalia: A Socio-Cultural Inquiry*. Ottawa: Canadian Government Publishing.

Woodward, R. (2005). From Military Geography to Militarism's Geographies: Disciplinary Engagements with the Geographies of Militarism and Military Activities. *Progress in Human Geography* 29 (6): 718–740.

Woodward, R., & Duncanson, C. (Eds.) (2017). *The Palgrave International Handbook of Gender and the Military*. London: Palgrave Macmillan.

1 Truth in military studies: lessons for producers and consumers of research on the military[1]

David R. Segal
University of Maryland

One of the major debates in the social sciences during the past three decades has been whether societies have transitioned from a modern to a postmodern era (Jencks, 1986). One facet of this prism has been the question of whether armed forces, which have frequently been viewed as modernizing institutions, have been part of this social transformation (Moskos et al., 2000).

The concept of a post-truth society is frequently seen as an attribute of postmodernism and its critique of reality, truth, and reason. Science, which has been committed to the pursuit of truth through the modern era, is an obvious target of postmodernists. Although trust in science, unlike other institutions, has remained fairly stable, in 2018, only 27% of US adults had great confidence that scientists would act in the best interest of society (Funk & Kennedy, 2019).

One of the characteristics of the modern era was the increasing role of rationality and science in the policy process and in the reciprocal influence of policy on science. This has been especially true in national security affairs. However, in recent years we have seen a rejection of science and of the concept of truth in the policy process in several modern nations.

In 2018, the Consortium of Social Science Associations (COSSA) in the United States held a Science Policy Conference in Washington, and asked me to speak about social science and national security in the US, where we have had a willing clientele in the security community since World War I, when military psychologists developed the army's first large scale selection and classification tests to measure the aptitudes of draftees. I participated in a panel on social science applications to enhancing security, drawing on a presentation I had given a decade earlier to two Congressional committees (Segal, 2008).

Two of the three plenary sessions at the COSSA meeting dealt with postmodern themes, addressing issues of "Post Truth: Communicating Facts, Not Fiction", and "Reestablishing Trust in Social Science & Data". This caused me to reflect both on the products that we provide to the policy clientele and how science has been used by this community, especially in the areas of diversity and cohesion. Issues of truth in the production and use of

DOI: 10.4324/9781003165217-1

social science research, both by policymakers and others, have been part of the process since World War II. In this chapter, I discuss lessons learned from the reporting and application of research in these areas.

The cohesion theme

Three World War II research efforts achieved iconic status both within the social sciences and the national security policy community; each has lessons worthy of remembering. The first is S.L.A. Marshall's *Men Against Fire* (Marshall, 1947), which for four decades following World War II was the most frequently cited study of combat in that war. Marshall reported, on the basis of about 400 group interviews with infantry companies conducted after combat, that no more than 15–25% of soldiers had fired their weapons in the direction of the enemy, denoting the so-called "ratio of fire". However, in a number of subsequent presentations, both the percentages of soldiers firing their weapons and the number of infantry companies interviewed varied. SLAM, as S.L.A. Marshall was called, was a journalist by training and vocation and frequently expressed a dislike for statistics. It may be unfair to hold him to the standards of social science, even though this is how he was regarded and initially embraced by social scientists.

In the postwar years, Marshall's work was widely cited, and his influence was not limited to the ratio of fire. When the racial integration of the US Army was discussed prior to the Korean War, and when gender integration of the army was debated in the early years of America's all volunteer force, the concept of cohesion was used to resist diversity on the basis of the assumption that social homogeneity produces cohesion. Marshall's work was widely cited with reference to the importance of cohesion (Segal & Kestnbaum, 2002). Marshall (1947, p. 42) had written that "…the thing that enables an infantry soldier to keep going…is the near presence or presumed presence of a comrade".

By the late 1980s, questions about the veracity of his findings were being published, as some World War II veterans reported gaps between their experiences and Marshall's reporting (Halloran, 1989; Smoler, 1989). Elements of his biography were also called into question. This might have been one of the first examples of a postmodern rejection regarding accepted wisdom. The controversy became so painful to the Marshall family that Marshall's grandson took a year off from his employment as a journalist – a vocation in which he followed in his grandfather's footsteps – to research the truth (Marshall, 1993).

The younger Marshall's research suggested that the broad strokes of SLAM's narrative held true. However, some important details such as the 15–25% ratio of fire figures and the number of companies interviewed, lacked support, and some biographical claims were untrue. SLAM still had his supporters in the army, and they characterized his misstatements of fact as "fibs" rather than lies, the distinction being that he reported things as he wanted them to be, rather than intending to deceive. Today, these might be called "alternate truths".

The esteem in which he had been held by military historians declined. The SLA Marshall Chair at the Army Research Institute, which had been established in 1986, was abolished. In the 1990s and the first decades of the twenty-first century, when policies regarding military sexual orientation integration and gender integration of ground combat units were debated, SLAM's work had largely disappeared from the discussions of cohesion.

Lesson 1. *If you want your research to have lasting relevance, tell the truth about what you found and how you found it. It's the ethical thing to do and if you lie, you may eventually be caught.*

The second piece of iconic World War II research from which we draw lessons on interpretation and application is Shils and Janowitz's (1948) classic study of the *Wehrmacht*. This study suggested that it was interpersonal dynamics in military companies (primary group cohesion and leadership) rather than commitment to National Socialist ideology or ineffective allied propaganda – that accounted for the willingness of German soldiers to continue fighting after it was clear that Germany was losing the war. This paper has been cited more than 1,000 times and it continues to be cited, including more than 50 times a year for the past few years. Although Janowitz was not yet a sociologist at the time of his service (Segal, 2017b), Shils had been a sociology faculty member at the University of Chicago before the war, and Janowitz went back with him to Chicago as a doctoral student after demobilization. I cannot fault either the methods or the conceptualization of this paper. However, I do have concerns about the way it has been cited by social scientists.

A multitude of published studies refers to the "*surveys*" of German PWs (prisoners of war) conducted by Shils and Janowitz. In fact, as the authors noted, the Psychological Warfare Division of SHAEF did conduct opinion polls of PWs to measure the effectiveness of propaganda and this was the subject of the only table in the article. However, their cohesion analysis was based on front-line interrogation of PWs and intensive psychological interviews in rear areas. Shils and Janowitz served with the Office of Special Services (OSS), a forerunner of the CIA, and their data were primarily intelligence interrogations. Many of them were part of the denazification process aimed at determining the order in which PWs would be released. The authors were explicit in admitting the lack of scientific rigor in their data collection. Citing the data from these interrogations as the results of sociological surveys suggests a lack of sensitivity to the difference between interrogations and surveys. The methodological issues raised by the former include reliability issues that may be very different from those raised by surveys. Under current American regulations regarding human subject research, PWs are a protected population who cannot be used as research subjects. It is possible, of course, that the authors of some of the papers that have cited Shils and Janowitz have not read the original text carefully (if at all).

Lesson 2. *Read carefully and report honestly the research methods used in the research upon which you build.*

The third iconic World War II research effort was reported in the four-volume series by Stouffer and his colleagues on *Social Psychology in World War II,* including the two volumes titled *The American Soldier* (Stouffer 1949a, 1949b).

Just as Shils, Samuel A. Stouffer had been a sociology professor at the University of Chicago before the war (Segal, 2017a). By the dawning of World War II, the army had realized that the attitudes of soldiers, such as morale, were as important as the aptitudes that had been studied since World War I in understanding military performance and effectiveness. Stouffer was asked to establish a research organization to contribute to such understanding; he recruited many of the major sociologists and social psychologists of his generation to develop a survey research program (Ryan, 2013). During the course of the war, the Information and Education Division's Research Branch, which he headed, surveyed more than half a million American soldiers, and reported their findings regularly to senior army commanders. Following WWII, they published their four-volume series with foundation support.

In 1988, I was teaching a course on Armed Forces & Society at West Point. At the beginning of our discussion on the issue of cohesion, two of my cadets told me that they had learned in their Combat Leadership course that Stouffer had found that cohesion was the major factor that kept soldiers going in combat. I had repeatedly heard this assertion in policy discussions when cohesion was being used to argue against diversity in the army.

I told these young cadets to read Stouffer et al. (1949b), to determine (a) how Stouffer's team measured cohesion; (b) whether in fact cohesion was the major combat incentive for soldiers; (c) what percentage of soldiers said that cohesion was the main factor; and (d) whether Stouffer's data provided a basis for generalizing to the army. I also asked them to share what they found with their Combat Leadership instructor.

They found, to their surprise, that cohesion did not appear in the index of the *American Soldier* volumes, but that the researchers had asked a question that had face validity in getting at combat motivation. *Generally, from your combat experience, what was most important to you in making you want to keep going and do as well as you could?* They were interested to learn that the research team surveyed army units, but did no army-wide surveys and that frequently the results from two or more units were combined in one table. Nowadays, this would be regarded as meta-analysis, with protocols for such data aggregation that did not exist in the 1940s. The cadets also noted that sometimes there were subtle changes in question wording, which is not surprising, since the team was exploring how to do better surveys while they carried out their research. And the cadets picked up on the fact that officers were often asked different questions than enlisted personnel to measure the same variable. These issues had relevance for the generalizability of the data.

With regard to the primacy of cohesion as a combat incentive, the data presented (Stouffer et al., 1949b), based on a survey of a division that had been in combat in North Africa and Sicily, showed "Ending the task" to be the most commonly cited factor among enlisted personnel (39%). "Solidarity with the group" (cohesion) was the second most important but at a lower level (14%) than one might expect if one thought that cohesion was the primary driver. Among officers in the same unit, "Leadership and discipline" had been the most important (19% but selected only by 1% of enlisted personnel), and "Solidarity with the group" was second (15%). The figures differed in other divisions, and cohesion generally ranked very high, but it never exceeded 20% of soldiers and was rarely the most important factor.

My Combat Leadership colleague came to see me after our cadets had reported their findings to him, and he told me that I had been right and that he had learned something. I learned something too, which I regard as the third lesson of this exercise, and a corollary of lesson 2.

Lesson 3. *When building on the work of earlier researchers, determine what the norms for conducting and reporting research were at the time they were doing their work.*

The racial integration theme

Another theme that appeared in Stouffer's army surveys was race relations. Before the war, the Carnegie Foundation had funded Swedish economist Gunnar Myrdal's (1944) study of race relations in the United States, and Myrdal had recruited Stouffer to assist him. When there was a threat of German invasion of Sweden, Myrdal returned home, and Stouffer completed the project. Stouffer saw the racial problems in the army reflecting the tensions that Myrdal had identified in American society more broadly.

The army during World War II was racially segregated, with a limit on how many Black soldiers could serve. However, late in the war, faced with a shortage of personnel to serve in infantry divisions, the army experimented with allowing Black soldiers to volunteer to serve in Black platoons that were incorporated into White infantry companies. Most of the daily life of infantry soldiers takes place within companies, and this composite organization reflected a step toward integration. Stouffer's research showed that acceptance of integration on the part of both Black and White soldiers increased with degree of proximity. However, at the end of the war, the army decided to assign Black soldiers to separate units, ranging from companies to regiments, some of which would serve in larger White units. The abolition of Black platoons in White companies decreased proximity. It changed Black exclusion from most infantry divisions to racial segregation within them. Moreover, President Truman made an informal agreement with the army that if the new system didn't work, the army could return to the old system of segregation (Binkin & Eitelberg, 1982, p. 28).

In July 1948, Truman issued Executive Order 9981, which called for equality of treatment on the basis of race, color, religion, and national origin. It did not require desegregation or integration. Under the legal doctrine of separate but equal that had been established in 1896 by the US Supreme Court, it provided support for opponents of integration. In response to the Executive Order, the army, in early 1950, did issue a regulation establishing "equality of treatment and opportunity for all personnel... without regard to race, color, religion, or national origin". This regulation could be used to justify either integration or segregation (Bogart, 1969, pp. 18–19). It did abolish the limit on how many Black soldiers could serve.

The Research Branch findings on race relations in the army were among its most frequently cited results and had two important spin-offs. For one, they helped shape motion picture producer Frank Capra's film, *The Negro Soldier*, which was part of his *Why We Fight* propaganda series. Interestingly, it did not reflect the more negative attitudes expressed in surveys by Black soldiers. This was an example of media license in the use of research: part of the complex dynamic of interpretation noted in the introduction to this volume. Also, the results regarding the effects of interaction between Black and White soldiers served as a basis for Gordon Allport's (1954) "contact hypothesis", published after the war in *The Nature of Prejudice*. In the long run, they may have contributed to the racial integration of the army, but did not do so immediately. However, both the army and the Democratic Party declared that Truman's executive order had integrated the army.

Many behavioral and social scientists embraced the official policy definition. For example, prior to the end of military conscription in the United States, Moskos (1966, p. 132) wrote "On July 28, 1948, President Truman issued an executive order abolishing racial segregation in the armed forces of the United States". Indeed, at an academic exercise at West Point in 2018, a cadet panel working with a behavioral science faculty member reported that President Truman had integrated the army in 1948. Many social scientists who use the 1948 date cite Moskos as the source. This commonly held assumption was also adopted by researchers studying other areas of diversity in the military who used the 1948 racial integration as their presumptive model (for example, McGraw et al., 2016, p. 8).

Whatever the interpretation of Truman's Executive Order, when the United States deployed army units to South Korea in 1950 in response to the North Korean invasion, these units were still racially segregated, both those that had been occupation troops in Japan and were the closest

American troops to Korea and others, which were deployed from the United States (Bogart, 1969, p. 49).

Lesson 4. *Do not assume that statements of official policy reflect the facts on the ground. Focus on the reality of the situation.*

The US Army during the Korean War was manned primarily by the small cohorts that were born during the Great Depression: even a large percentage of mobilization produced a relatively small army (Segal & Segal, 2004, pp. 4–5). The army was under great pressure to maximize the efficiency of the personnel it had. Building on its positive experiences with applied behavioral science in the two World Wars, it sought to continue this utilization. In March 1951, the Operations Research Office of Johns Hopkins University was tasked by the army "to initiate a project to determine how best to utilize Negro personnel within the Army". This research, designated *Project Clear*, was classified as SECRET from the outset (Bogart, 1969). The data were ultimately destroyed by the army, reflecting the military's control of information noted in the introduction to this volume.

The project director for *Clear* was Leo Bogart, a sociologist who, like Morris Janowitz, had served in military intelligence during World War II after graduating from college, credited his military experience with his attraction to sociology, and pursued a Ph.D. at the University of Chicago after demobilization (Bogart, 2004). Unlike Janowitz, he spent most of his career in marketing, advertising, and opinion polling, and was accustomed to proprietary research. However, from the outset, the scientists working on *Clear* felt that the project had important potential civilian applications. They attempted to get it declassified, which ultimately occurred in 1966, during the Vietnam War.

While some army leaders tried to maintain racial segregation early in the Korean War, units in the Far East began to be integrated. Some African-American replacements were assigned to White units and, in a few cases, White replacements were assigned to Black units. These changes reflected a number of factors, including the Far East Command honoring army policy to assign personnel without regard to race. In addition, with limitations on how many Black soldiers could serve having been lifted, Black replacements were arriving in excess of the personnel needs of Black units, while unfilled vacancies existed in White units due to combat casualties and fatalities. In some cases, assignments were made without awareness of the race of the soldier. As had been the case in Stouffer's research, White soldiers who served in integrated units were more favorable to integration than those who served in all White units. Indeed, a majority of the former soldiers surveyed favored assignment regardless of race.

Data collected from soldiers in Korea showed that, in mixed race units, integration had proceeded smoothly and without conflict. There was a general agreement that integrated units performed better than segregated Black units. The integration process continued through the war. In 1954, the US Supreme

Court rejected the separate but equal doctrine. The last Black army unit was deactivated that year and the Secretary of the Army announced that the army was racially integrated. However, it has been Department of Defense policy to continue to view 1948 as the date of racial integration of the American military.

The gender integration theme

The all-volunteer force established in 1973 would not have succeeded without an over-representation of Black soldiers. The other demographic group that contributed to the success of the all-volunteer force was women. During most of the period that the force was racially segregated, women were almost completely excluded. Until the start of the twentieth century, nursing services were provided to the army and navy by civilians under contract. Early in the twentieth century, nurse corps were established in both services, to be staffed by White women who were commissioned into special military branches and, thus, gender-segregated. In World War II, the exclusion of women who were not nurses ended with the establishment of gender-segregated auxiliary women's branches: Women's Auxiliary Army Corps (WAAC), the navy's WAVES (Women Accepted for Voluntary Emergency Service), and the Women Airforce Service Pilots (WASPS). The WASPs were uniformed, subject to military discipline, accepted the risks of flying, but served as civilians and did not have military status or benefits. Except for historical records, there was no research on the exclusion or acceptance of military women (Treadwell, 1954).

At the end of the war, authorization for the service of women was going to lapse. However, in 1947, World War II conscription also lapsed and, in order to avoid personnel shortfalls, the newly unified Defense Department sought authorization to have women serve in the regular force. The Women's Armed Service Integration Act, passed in 1948, provided military status for women, at a maximum ceiling of 2% of the enlisted force with women officers (not including nurses) authorized at 10% of enlisted women strength. Women in the air force and navy were not segregated in a separate branch, but were excluded from combat aircraft and from navy ships other than hospital ships and transports. Women in the army, other than those in medical specialties, were assigned to the gender-segregated Women's Army Corps (WAC). They were not excluded by statute from ground combat, but there were no combat specialties in the WAC.

The 1960s saw high levels of social turbulence in America, manifested in part by opposition to the Vietnam War and to the conscription system that provided its personnel, and by demands for equal rights for women and minorities. Partly in response to these factors, the 1948 Act was amended in 1966, removing the 2% limit on women's enlisted strength. However, the new legislation did not create gender equality of service conditions any more than Truman's Executive Order had ended racial segregation. Instead, it left army women in a separate branch, and it left separate promotion systems for women in the army, navy, and marine corps intact. It continued to perpetuate the exclusion of women from the service academies and the Reserve Officer Training Corps at civilian universities.

By the late 1960s, it seemed likely that military conscription was to be one of the casualties of the Vietnam War and much social research was conducted on the feasibility of an all-volunteer force (Tax, 1967). This research arena was dominated by labor economists, although sociologists such as Morris Janowitz and Charles Moskos also weighed in. The President's Commission on an All-Volunteer Force (Gates et al., 1970) made no mention of increased use of women in the military and seemed to believe that women would not have a significant role in the force. However, by the time the Congress passed the Equal Rights Amendment to the Constitution in 1972 (which was never ratified by the states), the military services were planning on promoting major increases in women's participation.

During the first eight years of the volunteer force, there was a quadrupling of the representation of women in the ranks, from 2% to 8%. In addition, there was an increase in the positions open to women, as well as important organizational changes. In 1970, the Reserve Officers Training Corps – the major source of officer accessions – was opened to women. In 1976, under a Congressional mandate, the military academies were opened to women: a phenomenon researched in all of the services, most extensively at West Point (Stiehm, 1981, pp. 118–145). Today, about a quarter of the enlisted ranks in the active force and somewhat higher percentages in the reserves are women. The expanded numbers and roles of military women generated, and were influenced by, a large body of social research, two elements of which I want to mention. The first is the fact that even before the 1978 dissolution of the Women's Army Corps, women were being integrated in regular army units. It was assumed that there was a ceiling on how many women could be accommodated without impacting negatively the ability of a unit to perform its mission. The same concern had been raised about the racial integration of units two decades earlier. The army conducted a series of three-day field exercises to determine the Maximum Women's Army Content (MAX WAC), varying and raising the percentage of women in 40 units up to 35% (U.S. Army Research Institute, 1977). The evaluation of these exercises showed that gender composition had no effect on unit performance. This was a classic example of begging the question: basing a study on an unsubstantiated assumption about a phenomenon and then trying to specify the parameters of the phenomenon. A more reasonable approach would have been to design research to answer the question "Is there a point at which changing the gender composition of a unit affects the performance of the unit".

Lesson 5. *Do not assume the outcome you seek in research. Ask the question you want answered.*

Finding no negative effects of increasing the percentage of women, the army voiced concern that a three-day field exercise in the US did not sufficiently approximate the conditions of an actual deployment. In 1977 women were included in the annual NATO REFORGER (Return of Forces to Germany) exercise, and the effects of gender integration on a ten-day deployment overseas were studied (Johnson et al., 1978). Once again, gender composition was not found to have an impact on unit effectiveness. In the intervening decades, no research has shown that the presence of qualified women has a negative impact on unit performance.

Increasing women's representation in the force continued through the 1970s, and was planned to increase through 1985. However, at the beginning of 1981, the services requested that the Reagan administration reexamine the expansion goals and keep the number of women at the 1980 levels (Segal, 1983, p. 203). The senior army leadership had suggested that the increased numbers of women in the army had contributed to a decrease in unit performance and combat readiness. On the other hand, Lawrence J. Korb, a political scientist who was an assistant secretary of defense, suggested that the "pause" in the expansion of women in the military that was established by the administration did not constitute a break in the trend of expanding women's roles in the military and said that allegations that combat readiness had suffered because of women were ridiculous and that "without women it would be lower than it is" (CQ Researcher, 1981).

There was an alternate potential explanation for the decrease in quality and performance being noted by army commanders. In 1976, as the representation of women was increasing, the Department of Defense started using a new selection and classification test, the Armed Services Vocational Aptitude Battery [ASVAB 76] to replace the tests then in use by the services. To maintain consistency over time, the scores of four subtests in the battery were combined into an Armed Forces Qualification Test (AFQT) score to be calibrated to previous tests so that percentile scores would be standardized for the post-World War II period. This test was the lineal descendent of the test program that had begun in World War I.

However, calibration to the earlier tests was erroneous. From the outset, the services were enlisting personnel whose scores grossly overestimated their cognitive abilities (Laurence & Ramsberger, 1981). Between 1976 and 1980, the services enlisted almost 425,000 Category IV recruits – the lowest acceptable mental category. This was about a quarter of new enlisted accessions. These were mostly men and about a quarter were assigned to combat jobs. Almost 360,000 young men who under correct standards would not have been eligible to serve were accepted (Eitelberg, 1988, pp. 172–182). It is no wonder that commanders in the field complained of poor performance. This had nothing to do with the gender composition of the force. The aptitudes of the women recruited during this period exceeded those of the men.

The Department of Defense recognized that there was a problem and committed itself to identifying and correcting it. Although it was not about gender, this is the second element of research affecting the utilization of military women that I want to mention. A number of new psychometric studies were commissioned (e.g., Maier & Truss, 1983; Sims, 1978). And a new test was instituted in 1980.

Lesson 6. *An important step in the imputation of causality is to consider the possibility of alternative causes.*

The sexual orientation theme

During the 1992 American presidential election campaign, candidate Bill Clinton promised that he would end the longstanding ban on homosexuals serving in the United States armed forces. While the American people were becoming increasingly tolerant of diversity in sexual orientation, the senior military leaders were opposed to that stance. The lesson that the military leadership had learned from the Korean War and *Project Clear* was that they should not ask questions if they would not like the answers. The Department of Defense directed the services not to conduct research on sexual orientation integration (Gade et al., 1996, 106). However, academic social scientists were beginning to turn their attention to this topic (e.g., Scott & Stanley, 1994) and in 1991, at a conference on Armed Forces in the Post Cold-War World organized by Charles Moskos and John Williams, one of the dimensions identified for discussion was that of homosexuals in the military.

Also, in 1993, the Army Research Institute, following the letter but not the spirit of the Department of Defense directive, asked British scholar Gwyn Harries-Jenkins to convene a meeting of European military sociologists in England to discuss the ways their countries were approaching sexual orientation issues in the military (Segal et al., 1993). The US General Accounting Office and RAND also published reports on homosexuals in foreign military forces.

The Army Research Institute asked me to attend the meeting in England. Subsequently, I was asked to testify to a US Senate panel on the policies and practices of our allies, as was Charles Moskos. Moskos and I agreed for the most part on the facts of what was policy and practice in the militaries we discussed. Drawing on the comparative data, he was more comfortable than I with the compromise Don't Ask, Don't Tell (DADT) policy that emerged from the process, and he was more critical of the applicability of the experience of nations that have less expeditionary forces than the US. We both noted that judgments on the issue of homosexuals in the military were in part subjective and not entirely data based.

Lesson 7. *In making policy recommendations on the basis of social research, it is important to distinguish between data and opinion. There may not be* *alternative truths but there may be alternative opinions on what the data mean which should be acknowledged.*

15 years later, the Department of Defense responded differently to President Obama's desire to repeal DADT than it had to President Clinton. It appointed a Comprehensive Review Working Group which, with a great deal of behavioral science input, provided a rigorous fact-based assessment of the impact repeal would have (Lee, 2013). This formative evaluation included a survey of more than 115,000 military personnel to assess the anticipated impact of repeal, if it occurred, on readiness, effectiveness, cohesion, recruiting, and retention (Segal et al., 2015). Despite the general pattern of resistance to change found in human populations (e.g., Hoffer, 1963), strong majorities of soldiers supported the repeal.

The Department of Defense did not conduct a post-repeal survey. However, a team of nine scholars – including several who had advised the Comprehensive Review Working Group, and behavioral science faculty members from West Point, the Naval Academy, the Air Force Academy, and the Marine Corps War College – assessed the impact of repeal one year after DADT was lifted (Belkin et al., 2015). The assessment involved surveys, interviews, and observations of units. The major finding of the assessment was that repeal of DADT was a non-event. The negative effects on readiness, recruiting, retention, cohesion, and morale that had been anticipated by some did not occur. All of the factors that had been attributed to the successful integration of African-Americans and women in the past were at work with the successful repeal of DADT. This result is consistent with what has been found in other western industrial democracies.

Lesson 8. *Despite differences between services, missions, and cultures, cross-national comparisons of military personnel policies and practices can provide useful baselines.*

Coda

Truth in military social science has been an issue at least since World War II and is not to be taken as an indicator that armed forces have become postmodern. It is not and they have not (Booth et al., 2001). Social scientists who study the military must be aware that in our pursuit of truth, while our parent disciplines are our quality control mechanisms, our field has always had an applied orientation and an outside constituency. Our serving these two audiences requires that we remain mindful of the fact that the product we provide at any given time is not absolute truth, but the best approximation we can arrive at given current methods, theories, paradigms, and assumptions, which we should make explicit. In this era of declining trust in institutions, the military, at least in the United States, is the only institution in which trust has increased (Funk & Kennedy, 2019), and social scientists who study the military may benefit from spill-over of this trust. However, we must also remember that we are not the policy-makers, who can use, disregard, or distort our findings.

Our contribution is to provide the best knowledge we can, recognize that we are only one source of input to the policy process, and hope that the policy-makers and others who use our data recognize the importance of what science has to contribute.

Note

1 This chapter is based on a keynote address presented at the 15th Biennial Conference of the European Research Group on Military and Society (ERRG-OMAS), Lisbon Portugal, June 17–21, 2019. I am indebted to Mady Wechsler Segal and John Allen Williams for comments on an earlier draft.

References

Allport, G. W. (1954). *The Nature of Prejudice*. Reading, MA: Addison-Wesley.

Belkin, A., Ender, M. G., Frank, N., Furia, S. R., Lucas, G., Packard, G., & Segal, D. R. (2015). Readiness and DADT Repeal: Has the New Policy of Open Service Undermined the Military? *Armed Forces & Society* 39: 587–601.

Binkin, M., & Eitelberg, M. J. (1982). *Blacks and the Military*. Washington, DC: Brookings.

Bogart, L. (2004). *How I Earned The Ruptured Duck: From Brooklyn to Berchtesgaden In World War II*. Chicago, IL: Markham.

Bogart, L. (Ed.) (1969). *Social Research and the Desegregation of the U.S. Army*. Chicago, IL: Markham.

Booth, B., Kestnbaum, M., & Segal, D. R. (2001). Postmodernity and the Modern Military. *Armed Forces & Society* 27: 319–342.

CQ Researcher. (1981). *Women in the Military*. Washington, DC: CQ Press.

Eitelberg, M. J. (1988). *Manpower for Military Occupations*. Washington, DC: Office of the Assistant Secretary of Defense [Force Management and Personnel].

Funk, C., & Kennedy, B. (2019). *Public Confidence in Scientists Has Remained Stable for Decades*. Washington, DC: Pew Research Center.

Gade, P. A., Segal, D. R., & Johnson, E. M. (1996). The Experience of Foreign Militaries. In G. M. Herek, J. B. Jobe, & R. M. Carney (Eds.), *Out in Force* (pp. 106–130). Chicago, IL: University of Chicago Press.

Gates, T. S. Jr., Curtis, T., Dent, F., Friedman, M., Greenwalt, C., Greenspan, A., & Wilkins, R. (1970). *Report of the President's Commission on an All-volunteer Armed Force*. Washington, DC: U.S. Government Printing Office.

Halloran, R. (1989, Feb. 19). Pivotal S.L.A: Marshall book on warfare assailed as false. *New York Times*.

Hoffer, E. (1963). *The Ordeal of Change*. New York, NY: Harper & Row.

Jencks, C. (1986). *What is Postmodernism?* New York, NY: St. Martins.

Johnson, C. D., Cory, B. H., Day, R. W., & Oliver, L. W. (1978). *Women Content in the Army-REFORGER 77 (REF WAC 77)*. Arlington, VA: Army Research Institute.

Laurence, J. H., & Ramsberger, P. F. (1981). *Low Aptitude Men In The Military*. New York, NY: Praeger.

Lee, J. L. (2013). The Comprehensive Review Working Group and Don't Ask Don't Tell Repeal at the Department of Defense. *Journal of Homosexuality* 60: 282–311.

Marshall, J. D. (1993). *Reconciliation Road: A Family Odyssey*. Seattle, WA: University of Washington Press.

Marshall, S. L. A. (1947). *Men Against Fire: The Problem of Battle Command*. New York, NY: Morrow.

McGraw, K., Koehlmoos, T. P., & Ritchie, E. C. (2016). Women in Combat: Framing The Issues of Health and Research for America's Servicewomen. *Military Medicine* 181: 7–11.

Maier, M. H., & Truss, A. R. (1983). *Original Scaling of ASVAB Forms 5/6/7*. Alexandria, VA: Center for Naval Analyses.

Moskos, C. C., Jr. (1966). Racial Integration In The Armed Forces. *American Journal of Sociology* 72: 132–148.

Moskos, C. C., Williams, J. A., & Segal, D. R. (Eds.) (2000). *The Postmodern Military: Armed Forces after The Cold War*. New York, NY: Oxford University Press.

Myrdal, G. (1944). *An American Dilemma: The Negro Problem and American Democracy*. New York, NY: Harper.

Ryan, J. W. (2013). *Samuel Stouffer and the GI survey*. Knoxville, TN: University of Tennessee Press.

Scott, W., & Stanley, S. C. (Eds.) (1994). *Gays and Lesbians in the Military*. New York, NY: Aldine De Gruyter.

Segal, D. R. (2017a). Profile: Samuel A. Stouffer. *The Military Psychologist* 32 (1): 20–24.

Segal, D. R. (2017b). Profile: Morris Janowitz. *The Military Psychologist* 32 (3): 20–24.

Segal, David R. (2008, April 24). *The Role of the Social and Behavioral Sciences in National Security*. Testimony to the Subcommittee on Terrorism, Unconventional Threats and Capabilities, House of Representatives Committee on Armed Services, and Subcommittee on Research and Science Education, House of Representatives Committee on Science and Technology. Washington, DC.

Segal, D. R., Gade, P. A., & Johnson, E. M. (1993). Homosexuals in Western Armed Forces. *Society* 31: 37–42.

Segal, D. R., & Kestnbaum, M. (2002). Professional Closure in The Military Labor Market: A Critique of Pure Cohesion. In L. J. Matthews (Ed.), *The Future of The Army Profession* (pp. 441–458). New York, NY: McGraw-Hill.

Segal, D. R., & Segal, M. W. (2004). America's Military Population. *Population Bulletin* 59 (4): 3–40.

Segal, D. R., Segal, M. W., & Reed, B. J. (2015). Diversity and Citizenship in Modern Military Organization. *Turkish Journal of Sociology* 35: 43–61.

Segal, M. W. (1983). Women's Roles in the U.S. Armed Forces: An Evaluation of Evidence and Arguments for Policy Decisions. In R. K. Fullinwider (Ed.), *Conscripts and Volunteers: Military Requirements, Social Justice, and The All-volunteer Force* (pp. 200–213). Lanham, MD: Rowman & Allanheld.

Shils, E. A., & Janowitz, M. (1948). Cohesion and Disintegration in the *Wehrmacht* in World War II. *Public Opinion Quarterly*, 12: 280–315.

Sims, W. H. (1978). *An Analysis of The Normalization and Verification of the Armed Services Vocational Aptitude Battery (ASVAB) Forms 6 and & 7*. Arlington, VA: Center for Naval Analyses.

Smoler, F. (1989). The Secret of the Soldiers Who Didn't Shoot. *American Heritage* 40 (2): 36–46.

Stiehm, J. H. (1981). *Bring Me Men & Women*. Berkeley, CA: University of California Press.

Stouffer, S. A., Suchman, E. A., DeVinney, L. C., Star, S. A., & Williams, R. M., Jr. (1949a.) *Adjustment to Army Life: Vol. 1. Studies in Social Psychology in World War II: The American Soldier*. Princeton, NJ: Princeton University Press.

Stouffer, S. A., Lumsdaine, A. A., Lumsdaine, M. H., Williams, R. M., Jr., Smith, R. B., Janis, I. L., & Cottrell, L. S., Jr. (1949b). *Combat and Its Aftermath: Vol.2.Studies in Social Psychology in World War II: The American Soldier*. Princeton, NJ: Princeton University Press.

Tax, S., (Ed). (1967). *The Draft: A Handbook of Facts and Alternatives*. Chicago, IL: University of Chicago Press.

Treadwell, M. E. (1954). *The Women's Army Corps*. Washington, DC: U.S. Government Printing Office.

U.S. Army Research Institute. (1977). *Women Content in Units Force Development Test (MAX WAC)*. Arlington, VA: U.S. Army Research Institute.

2 Science and politics: on the political utilization of military sociological research

The German Case

Gerhard Kümmel and Martin Elbe[1]

A study is a study is a study?

Military sociology in Germany is not a particularly popular discipline. Within academia, there are only a handful of scientists who once in a while do some military sociological research. For the past decades, the picture has been more or less the same. Yet, there is some military sociological research in the country, whose center of gravity rests with the Bundeswehr Center of Military History and Social Sciences (ZMSBw) and its predecessor, the Bundeswehr Institute of Social Sciences (SOWI), established in the mid-1970s. Both institutes are inhouse research entities, implying that they are fully funded by the German Ministry of Defense, and thus, engaged in commissioned research. This, however, does not mean that research is thoroughly controlled and directed by the MoD, given the specific institutional arrangement of the institutes.

In the following, we will analyze the institutional framework of the inhouse research institutes against the background of a system theoretical approach. The science-policy nexus that we have here, and that may well be transferred to other cases, is perceived as a systematic coupling between politics and science and in which the two influence each other in a specific way. Thus, the analysis basically falls into two sections, the first of which is of a general theoretical nature (Sections Diagnoses and When politics and science meet), while the other looks at the empirical reality in Germany (Section Military sociology and the department of defense in Germany). The paper explores the demands addressed to social science by politics, the ways in which these demands are met by social research, and how the findings of this military sociological research find their ways into politics. It hence shows "the impact of social research on the military itself", as it has been termed in the introductory chapter by the editors, and illustrates how politics deals with these findings.

To begin with, in a speech at the Nato Defense College in Rome, Dieter Mahncke[2] once addressed the question of how politics deals with the studies commissioned by politics and classified studies in the following categories, namely studies which:

DOI: 10.4324/9781003165217-2

- are ignored and left aside unread;
- are used in the political struggle because they confirm and justify the policies of the commissioner of the study;
- are read, but kept secret, because the research findings are not welcome; and
- are published and have a certain influence (Mahncke, 1994, p. 24).

This typology could well be extended by taking up Ulrich Beck's and Wolfgang Bonß' proposal regarding types of studies whose non-publication is "the prerequisite for their use" (Beck & Bonß, 1984, p. 399). A prominent example of the use of military sociological research results as an instrument to achieve one's own political objectives from the recent past comes from the speech by the then Federal Minister of Defense Dr. med. Ursula Gertrud von der Leyen in the parliamentary debate on the future German defense budget on November 25, 2015:

> We have commissioned a representative survey of the population by the Bundeswehr Center for Military History and Social Sciences (…) which shows that half of the Germans are in favor of increasing defense expenditures: 51%. This is the highest percentage that has ever been measured. Last year it was 32% and in 2013 it was only 19%. What does that tell us? It tells us – bitterly enough – that people are now, of course, aware of the seriousness of the situation. Secondly, it also tells us that there is trust in the Bundeswehr (…). But it also shows that people perceive that there is a need to catch up in terms of material and equipment and that security needs proper financing to prepare for crisis situations. This is what the 2016 defense budget proposal represents. It is a good defense budget; because it represents the necessary, long-awaited turnaround after so many years of shrinking. The proposed budget of 34.3 billion Euros allows us to continue on the path of modernization of the Bundeswehr that we have embarked on together. Therefore, I ask you for your consent. (Bundesministerin der Verteidigung Ursula von der Leyen, 2015, p. 13672)

The fact that the 2016 defense budget proposal received the backing of the parliament may indicate that "research becomes politically effective" (Mackewitsch, 1999, p. 89), at least sometimes.[3] The way in which politicians deal with military-sociological research results can be very different and researching the influence of military sociological research findings on practical politics is very difficult. Mackewitsch (1999, p. 89) states that direct evidence and measuring such military sociological impact on politics are "only possible to a limited extent". To make matters worse, in a "political society" characterized by political communication, democratic publicity, and increasing complexity and confusion (Greven, 1999), social science research results can also take chaos-theoretical paths because their effects become more complex the more actors are involved in the public and political discourse (see Beck & Bonß, 1989, p. 25).

Nevertheless, we will try to provide an analytical-conceptual approach to our topic and will begin with diagnoses of the processes and developments that can currently be observed in the relationship between science, society, politics, and consulting. This is followed by the elaboration of a system-theoretically based model of the relationship between science and political practice, which understands this network of relationships in terms of a structural coupling. Finally, the article uses this model to examine the relationship between military sociology and politics in Germany.

Diagnoses

The social scientification of the world

In his lecture on "Science as a profession", Max Weber describes the intellectualization and rationalization of the world, which he perceives as "disenchanting the world". The dynamics of scientific progress are considered "the most important part" of this process. The "intellectualist rationalization through science and scientifically oriented technology" practically means there are no longer mysterious and unpredictable powers at work and that things in the world can be calculated. (Weber, 1917/19, p. 9)

Against this background, modern society has become a "knowledge society" (Stehr, 2001), characterized by "everyday social science" (Beck & Bonß, 1984, p. 383), whereby the findings of social scientific research are practically "stripped" of their social scientific character (Kühl, 2003, p. 81) and become "invisible" (Froese et al., 2014, p. 9). Social scientific interpretation has found its way into society and politics to a large extent and has diffused into political and everyday language. By this it has become self-evident and invisible, but at the same time, also relevant (Beck & Bonß, 1984, pp. 392–395). Our everyday conversations provide plenty of evidence that "scientific knowledge has trickled down into our living worlds (…) and has sometimes been trivialized to common sense". (Raphael, 2018, p. 14) The widespread use of terms such as 'role' or 'socialization' illustrates this "diffusion of sociological terms into public and political usage" (Kaufmann, 1976, p. 305). For Lutz Raphael, this represents the "scientification of the social" (Raphael, 2018, p. 14; see also Sala, 2017), even though such a reception of sociological concepts does not mean that the underlying social science problems are adequately understood. In that sense, "the reception takes place selectively". (Kaufmann, 1976, p. 305)

The scientification of politics

In the course of the process of intellectualization and rationalization described by Weber, scientific knowledge and scientific expertise gained increasing relevance in the political sphere. This implies that the scientification of the social is matched by the scientification of politics. "Since our societies' self-image is democratic, since we accept no other power than the will of the people, science provides influential arguments about what is feasible and what is not". (Stengers, 1998, p. 7) Thus, in modern societies, the most influential argument is the one based on science (Stengers, 1998, p. 96). From the end of the Second World War to the beginning of the 1970s, there was an increasingly strong belief that political practice can be made better by the use of science (see Raphael, 2018). The underlying socio-technological understanding that scientific knowledge is "objective" is well documented in the concept of the "technical" or technocratic state (Schelsky, 1961, pp. 453–460). In this vein, politics likes to see society "as a machine with screws: if you turn the right screw, the world works as you wish. And the task of science is to label the screws correctly" (Streeck, 2009, p. 13).

This, however, evoked "the association of the technical, almost post-political" (Rehberg, 2001, pp. 1318f.) because of the implicit technocratic instrumentalization of research findings that covers its political character (Schneider, 1989, p. 326). This may well carry the risk of undermining democracy and replacing the essence of democracy, public debate, by technocratic socio-technological orders from science (see, e.g., Schelsky, 1961, p. 453). The crisis processes in the 1970s undermined the belief in the objectivity of scientific research findings not only in politics, but also in society. In this vein, there was talk of the "disenchanting of science" (Bonß & Hartmann, 1985), and a "reflected skepticism" concerning the findings of science emerged (Weingart, 2003, p. 10).

Nevertheless, politics still needs science; the scientification of politics and the "scientification of institutions and public discourses" are obviously processes that are irreversible (Beck & Bonß, 1989, p. 32). Accordingly, politicians regularly refer to scientific studies in their statements, so that there is a "mixture of science and politics in governmental action" (Murswieck, 1994b, p. 104). Scientific policy advice continues to have a certain relief function for politicians, as unpleasant or unwelcome decisions can simply be externalized to the consultants (Bolz, 2001, p. 19; see also Badura, 1980). Accordingly, a politician's staff is tasked with establishing a network in the respective political field and to find relevant actors, which may also include scientists who research aspects of this political field. From this group, "any politician (…) takes the scientific advice that is most likely to support his decision and gives the impression that the decision is nothing other than the compelling result of scientific knowledge" (Spaemann, 1990, p. 21). Finding counter-experts is often not particularly problematic, especially in the social sciences, since we can observe an increase and pluralization of the knowledge offered in the social sciences, which has provided the user of this knowledge with a certain autonomy of its use (Beck & Bonß, 1984, pp. 397–399).

Often, these experts are not only included in groups of experts or commissioned to prepare expert reports, but are also appointed to expert committees that are active over a longer period of time. However, since they thus exert a substantial influence on political decision-making (Schützeichel & Brüsemeister, 2004a, p. 7), the critics speak of a suspension of parliamentary democracy. Often, they also see scientific policy consulting as a "post-rationalization of practice", i.e., expert knowledge is needed for the justification of the policies already implemented (Bolz, 2001, p. 19). This function of obtaining ex post legitimacy for political decisions already taken is not always recognizable to the public.

The socialization of consulting

The skepticism toward science described above has led to the search for advice, guidance, and orientation in non-scientific areas. This also significantly increases the number of potential sources of advice and consulting actors. Since the 1970s, we have literally been living in an "age of therapy" (Raphael, 2018, p. 32; see also the articles in Maasen et al., 2011), because we can hardly do without consultations, guidance, advice, and counseling. Our society is therefore a "counselled society" (Schützeichel & Brüsemeister, 2004b) and a "McKinsey society" (Schützeichel & Brüsemeister, 2004a, p. 8).

Indeed, the consulting industry is a growth industry, and "[t]he guild of consultants has a bright future. (…) More and more people and organizations earn their money with consulting services, and more and more people and organizations will resort to consulting services" (Schützeichel & Brüsemeister, 2004a, p. 7). At the turn of the millennium, already 12–15% of sociologists graduating from university turned to the organizational and management consulting business (Saam & Petran, 2001, p. 7). Indeed, we seem to be in need of ubiquitous advice, for example on lifestyle issues, cooking, financial matters, career choices, consumer decisions, looking for a partner, raising our children, buying electrical equipment or cars, partnership problems, medication, insurances, and sexuality problems. Yet it also is the case when deciding on the establishment of no-driving zones for diesel vehicles in major German cities or when choosing a suitable location for investments, for example. Accordingly, the consulting business is booming and has not only caught the individual, but also collective actors such as companies, organizations, associations, and parties. The numerous so-called "influencers", who now find a mass audience on YouTube, are perhaps the most impressive illustrations of "how consulting has anchored itself in society" (Schützeichel & Brüsemeister, 2004a, p. 8; see also the contributions to Rietmann & Sawatzki, 2018, especially Engel et al., 2018). Thus, the consulting business has become more competitive and still remains of considerable importance.[4]

The politicization of science

With regard to the sciences, we can observe a politicization of science that matches the scientification of politics (Jann, 1994; Brückweh et al., 2012). As much as politics cannot evade its scientification, science cannot escape its politicization, even if it wanted to. According to Ulrich Beck (1980, p. 417), science has already been expelled from the ivory tower.

Science strives for knowledge, or according to Luhmann (1991), for truth. It produces knowledge and insights while being suspended of the need to act (Habermas, 1973, p. 214). Niklas Luhmann also emphasizes the 'action-free' reflection in science and sees it as the functional characteristic of the science system (see Luhmann, 1977, 1991). Often scientists want to leave it at that and to leave out the dimension of the practical application of their findings (Schneider, 1989, p. 302). In addition, science oftentimes values its autonomy and distance from politics (Mayntz, 1994, p. 17). Such an 'ivory tower' mentality, however, can hardly be sustained, since the codes of science themselves undermine this form of self-sufficiency in the pursuit of scientific progress. In order to gain recognition of one's own scientific work and its findings, these must be published in the usual scientific media. This means that the findings are public and can be accessed and used at any time by any member of the scientific system, but also by everyone outside the scientific system.

Frequently, however, the pursuit of gaining knowledge is explicitly linked to the search for solutions to real and practically existing problems. In recent years, against the background of growing ecological problems and in view of developments in the fields of biotechnology and genetic engineering, for example, there has been talk – in the sciences themselves, but also in parts of civil society and by some state governments – of the need for a transition to what has been termed 'transformative science' (see also Strohschneider, 2014). Under the heading of "Responsible Research and Innovation" (RRI), the European Commission has been aiming for a "new positioning of 'societal concerns' within research and development" in its research funding policy since 2013 in order to integrate "political and ethical perspectives in scientific, technical and medical research projects right from the beginning" (Borck et al., 2018, p. 215).

In the social sciences, this refers to questions about the good society, the good order, and about the proper ways to get there (Dahrendorf, 2001, p. 1332; Allmendinger, 2001). This sometimes persuades scientists to approach non-science actors and also politics, but also non-science actors from politics, business or civil society to approach science actors. With this in mind, in his Presidential Address at the American Sociological Association meeting in 2004, Michael Burawoy explicitly called for sociology to move toward a "public sociology" that actively and deliberately moves beyond the boundaries of sociology and addresses non-science actors (Burawoy, 2005).

With regard to this relationship to non-science actors in general and to politics in particular, there sometimes is the assumption of a rationality gap in favor of the sciences which may also lead to feelings of being somehow superior to them. What is needed instead is modesty on the part of the sciences (Streeck, 2009, p. 13). In addition, this points to the issue of the various individual professional understanding of scientists. In this vein, Roger Pielke (2007), for example, ideally distinguishes four types of scientists: (a) the 'pure scientist', to whom practice and politics do not matter; (b) the 'science arbiter', who makes his research available to the political-practical process without wanting to influence it further (c) the 'issue advocate', who is strongly advocating the practical implementation of his findings; and (d) the 'honest broker', who only explains to the practical-political decision-maker the range of options available. So while "issue advocates" try to prescribe a specific action option for politicians, the "honest brokers" convey to politicians a wider range of available options and thus give politicians greater scope for decision-making.

From a critical perspective, however, the variety of types of scientists is much less pronounced. According to this, policy advisors are nothing more than "courtesy scientists" (Bolz, 2001, p. 19). In this reading, scientific policy advice is determined by the specifications of the client, so that scientific policy advice represents a "conspiracy with the rulers", is a "dirty business" and should thus be refrained from (Offe, 1977, pp. 323f., pp. 328f.). This is even more valid as politics is tempting and seducing science. Social science graduates, e.g., who work in the consulting business may have to "dramatize their diagnoses (…) in order to turn the diagnosis into a political problem", which may then lead to access to research funds and/or follow-up contracts (Friedrichs et al., 1998, p. 22).

It is undisputed that the functional logic of science and politics are very different. Politics follows completely different codes, rules, and mechanisms than science and is oriented toward decision-making and action. This implies that time in politics ticks differently than time in science as has been finely described in literature by both Joseph Breitbach (1987) and Harry Mulisch (1992).

However, this does not have to be a reason to completely refrain from scientific policy advice, as long as one remains aware of the different system logics and functional logics of science and politics which should be a reminder not to indulge in undue delusions and illusions regarding the effects of political science research, but to be modest. Perhaps this is also to be understood as the task which is implicitly put to science by society who at the end of the day finances the sciences and provides the resources. In such a way, science has to accept being asked by society about the benefits it provides to society. The empirical reality actually provides evidence that "scientific knowledge has helped politics to make better decisions and to be generally able to do so" (Murswieck, 1994a, p. 12). Thus, there is still a prospect of the social sciences making politics more rational (Mielke, 1999, p. 18; Streeck, 2009, p. 13) following the enlightenment model described by the editors in their introduction.

This discourse is based on the power of evidence and the more appropriate argument in the fight of arguments, which reveals the accusation of "courtesy science" and "complicity with the rulers" as unsubstantiated. Scientific policy advice understood in such a sense is not the practically omnipotent royal science, but in the "political society" (Greven, 1999) we live in, a kind of scientific control loop for politics, which is by no means de-democratizing, but democratic in nature.

When politics and science meet

Scientific policy advice in general and social scientific policy advice in particular constitute a field in which the functional logics and autopoetic mechanisms of both politics and science meet. It can now be assumed that in this meeting area, the codes of both systems may no longer be as stringent as inside the respective system, but may rather, if only temporarily, experience a certain softening.

The social scientist is initially concerned with identifying social issues, adequately describing them and explaining them conclusively, while also trying to guarantee that the diagnosis adequately takes into account all relevant factors and context effects. The quality of the diagnosis determines the sustainability of the explanation of the problem, the quality of the answer to the respective research question, and the validity of the extracted forecast or the designed scenarios about future developments. When the duet of diagnosis and prognosis becomes a triad of diagnosis, prognosis, and action, and when scientific policy advice tries to inform and influence political decision-making and political action, the system of science moves toward the system of politics (Friedrichs et al., 1998).

But politics must also move. If one conceives scientific knowledge as an input for the system of politics, then the acceptance of this input is dependent on the "transfer affinity" (Froese et al., 2014, p. 2) and "on the connectivity in the receiving system" (Luhmann, 1991, p. 638). At the same time, this means that social science findings are rarely adopted and implemented unchanged by politics. Rather, these findings are modified and transformed on their way into the "user systems" (Luhmann, 1977, p. 18) of politics (Ronge, 1989, p. 332). As a rule, the practical use of scientific knowledge does not take the form of a pure, deductive application, but these scientific results are reformulated in the political process, inductively reproduced, and thus reconstructed. "In order to be practical, the arguments [of the social sciences] must rather be 'transformed', that is, stripped of their social-scientific identity and 'newly' constituted in accordance with the conditions of the practice" (Beck & Bonß, 1989, p. 11, quote p. 27). The transformation process of social science research results thus resembles a "construction by the user" (Luhmann, 1991, p. 638), which also includes the

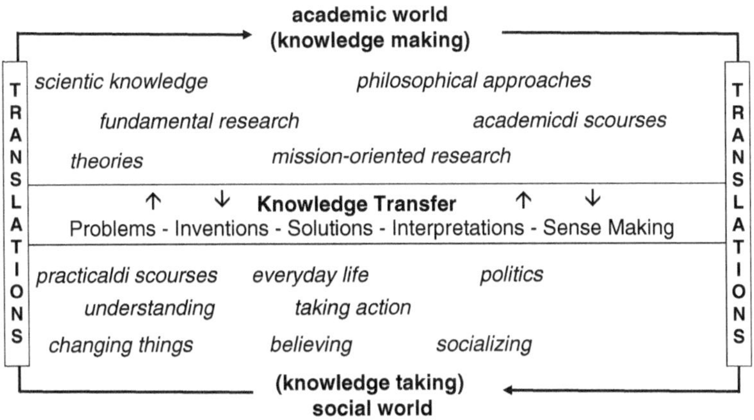

Figure 2.1 Process Model of Knowledge Transfer.

Source: Adapted and modified from Froese et al. (2014, p. 5).

"picking of results" by practical politics and the "trivialization" of social science knowledge in its use (Beck & Bonß, 1984, pp. 392–395).

The area of scientific policy advice can thus be understood as a structural coupling of the functional systems of science and politics (Luhmann, 1991, p. 22; Weingart, 2003, pp. 87ff.). The system-theoretical concept of structural coupling can explain "how autopoetic and self-referential systems (…) can come into contact with one another and co-develop in a coordinated manner" (Wiarda, 2015, pp. 31; see also Schluchter, 1971, pp. 56f.). The area in which social sciences and politics meet may be thought of as a revolving door that connects the two systems or as an osmotic area in which non-systemic elements protrude into the systems from outside. This exchange process may be explained by the concept of knowledge transfer (see Figure 2.1), which, however, is not to be understood as a linear and unidirectional movement, in which science passes on its research findings to various non-scientific actors in practice, here: to politics. This is an interactive and reciprocal process and includes feedback loops. Science transfer thus lies between the process of knowledge generation in the system of science and the process of the practical implementation of this knowledge in the system of politics (Froese et al., 2014, p. 4). With Wiarda (2015, Chapter 5), this knowledge transfer can also be viewed as a network that has system-like features.

Military sociology and the department of defense in Germany

If we now turn to the case of military sociology in Germany and its relationship to politics, the processes identified above, namely, the social scientification of the world, the scientification of politics, the socialization of consulting, the politicization of science, and the process model of knowledge transfer can very well be observed. Yet, some specifics are striking for the German case. One usually thinks of a policy adviser as an advisory agency, which employs a certain number of full-time advisors and always tries to attract orders to ensure the economic survival of the agency and its members (see Chahoud, 2010). Things are significantly different with regard to German military sociology in the Bundeswehr Center for Military History and Social Sciences (ZMSBw) and its predecessor, the former Bundeswehr Institute of Social Research (SOWI). In this case, we are dealing with an institutionalized body that was launched by the political addressee of military sociological policy advice, the MoD. Military sociological research thus takes place in an inhouse research body (see Mackewitsch, 1999; Klein, 2004; Zoll, 2016; Heinemann, 2016) by "'native' researchers formally linked to [the armed forces]", as put by the editors in the introductory chapter.

The creation of the SOWI was due to a favorable domestic and foreign policy constellation in the second half of the 1960s and early 1970s, when there were massive problems of recruiting volunteers, serious concerns about the acceptance of the Bundeswehr's basic organizational philosophy of "Innere Führung" among the troops, an increase, though small at the beginning, of the number of conscientious objectors; the threat of a nuclear war; the transition to deterrence as the primary task of the Bundeswehr; and ever louder calls for democratization and political participation by society. In this context, the MoD appointed, among other things, a commission on issues of the military's personnel structure in 1970 that proposed the introduction of compulsory academic training for officers in its 1971 final report.

Under the leadership of the Social Democratic Defense Minister Helmut Schmidt (1969–1972), the MoD commissioned the formulation of the curricula of the two Bundeswehr universities in Munich and in Hamburg to the military's Scientific Institute for Education in the Armed Forces, which moved from Heide to Munich in 1971 and was led by the political and administrative scientist Thomas Ellwein. Given the widespread belief of politicians in science (see above) at the time, this institute was transformed into the Bundeswehr Institute of Social Research in 1974. The rationale for doing so was laid out in the "White Paper 1971/72", which read: "The Bundeswehr and the public have an interest in the problems of national defense being examined scientifically" (Presse- und Informationsamt der Bundesregierung, 1971, p. 93). The political leadership of the MoD put through this idea against resistance in the military and in the ministerial apparatus who both "had their problems with an independently and critically working scientific institution" (Heinemann, 2016, p. 47).

The MoD itself justifies the establishment of the SOWI as follows:

> The establishment of the Bundeswehr Institute of Social Research was part of the intensive efforts to reform the Bundeswehr in the early 1970s. (...) The SOWI was tasked with providing, through independent social scientific research, insights into the social conditions and processes in the armed forces, into problems with organization and training, and in particular into the relationship between the Bundeswehr and society. (Jung, 1983)

The 1984 version of the SOWI's statute reaffirmed the character of the SOWI as a departmental research institution that carries out social research relevant to the military as commissioned research, but that should also be active in basic research (Bundesministerium der Verteidigung BMVg, 1984). It was oriented toward problem-solving from the very beginning. Brigadier General Jörg Sohst, then Head of Department in the Leadership Staff of the Bundeswehr, formulated the MoD's expectations with regard to the SOWI by stating "that military-related research in the social sciences should identify and analyze problems for the Bundeswehr and provide proposals for solving these problems" (Sohst, 1999, p. 73). SOWI Director Colonel i.G. Heinrich Geppert also perceived the SOWI as some sort of an "early warning system" for the armed forces and the MoD (Geppert, 1998, p. 21).

The research topics of the SOWI and, later on, the ZMSBw, naturally changed and expanded over time. For example, the SOWI's first statute wanted the SOWI to focus on the study of civil-military relations in Germany, problems of information and communication in the Bundeswehr, training and advanced training in the Bundeswehr, and the development of quantitative methods and processes in empirical social research. This also included basic research on these subject areas (Bundesministerium der Verteidigung BMVg, Führungsstab der Streitkräfte I 15, 1976). These areas of activity and research were supplemented in the Task Concept of 1997 by the topics of multinational cooperation between armed forces, international security and defense policy, and sociopolitical developments (Bundesministerium der Verteidigung BMVg, 1997; Geppert, 1998, p. 14). Later on, the SOWI studied the Bundeswehr's missions abroad; the integration of women into the Bundeswehr; issues of work-life balance; the integration of soldiers from the National People's Army of the GDR into the Bundeswehr; and issues of leadership, control, and personnel development (see the contributions in Dörfler-Dierken & Kümmel, 2016, Chapter IV; and the annual reports of the ZMSBw at http://www.zmsbw.de/html/publikationen/jahresberichtezmsbw).

The inhouse military sociological research of the SOWI and the ZMSBw and the reception of the military sociological research findings by the MoD are to be understood in the sense of the above-mentioned structural coupling between the functional systems of science and politics and can be understood as one multi-stage or multi-phase process (see Wiarda, 2015, pp. 36–39), whereby the structural coupling is already given by the fact that the ZMSBw is an inhouse research institution:

In Phase One, either the MoD or the ZMSBw sends a signal to the other system. For example, within the ZMSBw's research process, a problem may come up that needs to be dealt with by politics, or the MoD needs practical information from the ZMSBw and approaches the ZMSBw with a corresponding research request. This is done either through the official channels, i.e., through the institutionally prescribed paths, or informally and spontaneously. In both cases, this information trigger can develop a communication and coordination process. If the information trigger does not find any response from the other party, the process of re-activating the structural coupling is already over before it even started.

If the information trigger is taken up, Phase Two starts. Here, the two systems enter into a negotiation process, in which the parameters of the emerging research project are determined and the communication partners or negotiating partners agree on the goals, methods, procedures, and time horizons of the research. Ideally, this process is consensual. However, this is not always the case, due to the different quality of the arguments put forward. While the ZMSBw goes into the negotiations with a battery of scientific arguments, the MoD may simply rely on the resource of power and on the fact that the principal can simply instruct his agent to do something or not.

Phase Three starts right after that. The main actor in this third phase is the ZMSBw, since it is now either starting research or reorganizing existing knowledge from previous or current research. Sometimes a communication loop with the department is required in this phase, for example to obtain research-relevant practical information or if it turns out in the research process that the periods are too short or if there are interim results that require an intensification or expansion of the research efforts. This phase ends with the writing of the research report.

In Phase Four, the ZMSBw notifies the MoD of the results it has found, which is usually done via institutionally prescribed communication channels, but sometimes also through more informal channels. These results can vary in character. For one, they can be a purely informative service. This applies prominently, for example, to the ZMSBw's annual population surveys, in which the opinion of the German population on foreign, security, defense, and military policy issues is analyzed. In addition, the ZMSBw also provides consulting services for the department, i.e., the research findings are combined with the presentation of various options for action or with specific recommendations of action. In this context, the studies of the ZMSBw on personnel and personnel development questions, questions on military leadership, military career developments, and multinational cooperation should be mentioned.

In Phase Five, the results and/or recommendations of the ZMSBw are processed and converted into political decisions. The MoD may also request implementation services from the ZMSBw. This refers, for example, to the cooperation and participation of the respective scientists in intra-ministerial committees and steering groups or also in commissions that are installed on a specific topic and take practical decisions or at least significantly influence them. For example, the former SOWI was a member of the task group "Women in the Armed Forces", which was established after the judgment of the European Court of Justice in January 2000. Finally, transfer-based delegation services may also be requested by the department. In such a case, scientists from the ZMSBw work for a limited period of time in the MoD and bring in their expertise to a specific topic. In the recent past, this concerned, for example, the question of how to deal with rituals in the troops that violate the principles of internal leadership and have consequently triggered a social and political debate. In general, the ZMSBw's military-sociological policy advice is limited to information or advisory services; by contrast, implementation and delegation services are extremely rare. This also indicates that the structural coupling between military sociology and the MoD is very much influenced by procedural requirements and by provisions that regulate the relationship and cooperation between military sociology and the MoD.

What is missing so far is a Phase Six, in which the MoD discusses the decisions taken with the ZMSBw, or at least informs the ZMSBw about the decisions made.

In addition, the basic documents of the institute's work have undergone revisions and changes over time. For instance, critics at times perceived the institute as being on the way to its militarization and in danger of losing its scientific sovereignty (Hauer, 1999). The 1997 Task Concept explicitly oriented research tasks more toward the needs of the armed forces:

> However, the status between research for the defense department and academic research required for the SWInstBw must be oriented in such a way that the work of the institute is primarily based on the knowledge interests and performance benefits of the defense department. (...) The new tasks of the armed forces and the scarcity of social science research capacities must be taken into account by strengthening the focus of social science research on the needs of the armed forces. (Bundesministerium der Verteidigung BMVg, 1997, p. 2, p. 5)

This criticism was also fueled by the fact that after the relocation of the SOWI from Munich to Strausberg east of Berlin, for the first time an officer was appointed Director of the institute (Klein, 2004, pp. 40f.). In this context, however, it should be mentioned that a soldier can sometimes be more critical than a civilian. The underlying judgment that a military leadership is naturally a bad sign for a departmental research institution may not always be valid.

In contrast, two changes in the contents of the basic documents are more important: In its early years, the SOWI had "the right to invite journalists and hold press conferences independently" (Zoll, 1999, p. 126) in order to address the public. This constituted a considerable power tool, because "press conferences were the means to bring the research results into the media with daily updates. This not only achieved a broader acceptance of the results, but also created political pressure to take them into account in decision-making processes" (Zoll, 1999, p. 128; see also Bonacker, 2016, p. 119). The SOWI retained this right to a publicly recognized transfer service until the mid-1990s. Given that both the SOWI, and later on, the ZMSBw, are constrained to ex post transfer services, i.e., after the research results have been released to the public with the prior consent of the MoD. These transfer services usually come as publications and lectures.

A second significant change was implemented regarding the position of the Director of the institute. According to the 1974 Statute, the director was initially "directly subordinate to the Federal Minister of Defense and had the right to speak to the minister directly. The main communication was through the Parliamentary State Secretary" (Zoll, 1999, p. 126; see also Bonacker, 2016, p. 119). The SOWI inherited this right from its predecessor, the Scientific Institute for Education in the Armed Forces (Heinemann, 2016, p. 46), but later on lost it (Klein, 2004, pp. 40f.). The ZMSBw also does not have such a right (Zentrum für Militärgeschichte und Sozialwissenschaften der Bundeswehr ZMSBw, Kommandeur, 2017).

This in turn implies a strengthening of the 'apparatus'. Indeed, the structural coupling between the ZMSBw and the MoD does not consist of two, but rather of at least three actors. The MoD actually consists of two sub-actors, the apparatus and the management (ministry), who may have diverging and even opposite opinions. This constellation may become explosive: "The occasional controversy persisted and the fronts remained essentially the same: the social democratic leadership of the ministry supported the SOWI and its leader, the military apparatus and the CDU/CSU struggled with the critical scientists" (Heinemann, 2016, p. 48). Upon the abolition of the right to present to the minister or to the state secretary, an important formally established communication line ceased to exist. Since then, communication has been indirect via the bottleneck of the apparatus. Circumventing the apparatus was and is possible in extreme emergencies and by informal means only.

The knowledge transfer into the MoD and its leadership now needs more stations. Being familiar with the popular children's game "Silent Message", it is well known that the message from the sender rarely arrives at the final recipient in an unchanged and unfiltered way, so that the influence of the individual person on the communication process increases. It is very common, in terms of procedure, that the military sociological research results are summarized and condensed for the leadership by the responsible bodies in the apparatus. If the person responsible for this has an understanding that military sociological research essentially serves to protect the Bundeswehr's and the MoD's flanks (see Hagen, 2012, p. 41, footnote 24), it cannot be ruled out that this also has a certain influence on which findings find their way to the leadership and which findings don't.

Social research in the military is not a mere academic act, but part of micropolitics in the military organizations. Research results are not only the basis for consulting the MoD, giving advice may mean taking sides. Studies are used in the MoD to support goals of specific interest groups; e.g., to foster the program "Realignment of the Bundeswehr" was introduced in 2011. Secretary of State Stéphane Beemelmans presented a volume that reflected this process by the SOWI (Richter, 2012; Sozialwissenschaftliches Institut der Bundeswehr, 2012).

Against this background, SOWI employees have also seen the danger that military sociology may degenerate into mere social technology (Adorno et al., 1972; Habermas & Luhmann, 1985; Lippert & Wachtler, 1982, p. 353). In doing so, they reproduce the objection frequently voiced in the social sciences that "military sociology is often mentioned as a kind of applied company sociology, which works out the conditions under which the efficiency of the military organization can be maximized" (König, 1968, p. 9). The opposite and complementary suspicion on the side of the MoD can be found, for example, in the statements of Vice Admiral Hans Frank who, as current Deputy Inspector General of the Bundeswehr, voiced the view that the military should not be subjected to the SOWI's research in the same manner as a patient is subjected to the surgeon's scalpel (Frank, 1998, p. 4). Also Brigadier General Jörg Sohst called for an understanding on the side of the researchers that the MoD should be informed about research findings prior to the media and other actors (Sohst, 1999, p. 73). The client of military sociological research would like to remain "the master of the procedure" and "be able to readjust" it (Mackewitsch, 1999, p. 87). Accordingly, the ambivalence of the co-existence of a rhetoric of trust on the one hand, and an action logic of distrust on the other, continues to be a hallmark of the structural coupling of military sociology and the MoD.

What to do?

Of course, there is always the risk that the codes of the MoD try to intervene in military sociology, sometimes more, sometimes less. However, it would be exaggerated to speak of a thorough subjection of military sociology to the MoD's command, because the "barriers that separate science and practice (…) ensure (…) the identity and autonomy of science as a condition of its attractiveness for potential users. They offer a minimum level of protection against interventions by practitioners. The right balance has to be found, which allows for distinguishing between legitimate interests of use and abusive attempts at instrumentalization" (Schneider, 1989, pp. 302f.). The MoD must practically allow military sociology to follow its own codes to an extent that has to be constantly re-balanced so that it can produce results that are valid and "epistemically robust" within its own reference framework (Froese et al., 2014, p. 4), because only in this way can they become good products of scientific policy advice. One indicator for such an enlightened self-interest on the side of the MoD is that it still sticks with the research institute (Schubert, 1983, p. 28).

Further on the current conjuncture, ZMSBw offers more favorable conditions than the previous SOWI because the orientation toward basic research and the autonomy emphasized by the Science Council in the evaluation of the SOWI (Wissenschaftsrat, 2009, see also Bonacker, 2016) is much more pronounced here. However, it must also be stated that the unique position of military sociological research at SOWI/ZMSBw has been lost to some extent and that the competition for advice from other actors has increased somewhat. Military sociology, every now and then, will feel tempted to serve the MoD for the purpose of "a participation in rule" (Nullmeier, 2011, p. 42; see also Wiarda, 2015, p. 284) or to sell the MoD its product of scientific policy advice in the fashion of an "issue advocate" as the only adequate solution to the problem given, so that military sociology needs constant self-reflection, a pragmatic attitude and modesty (see Brodocz & Noetzel, 2004, p. 208; Buchholz, 2008, p. 17; Streeck, 2009, pp. 14f.; Rüb, 2011).

In order to find a balance that is mutually satisfactory, it is important to productively overcome "sterile camp thinking" on both sides (Bredow, 1999, p. 48). Distrust can only be reduced if trust is built. This requires continuous communication and a systematic exchange between research and practice, i.e., practically a continuation of the structural coupling in the sense of a co-evolutionary development and learning processes (Howaldt, 2008, pp. 34f.; Luhmann, 2005, p. 375). However, one has to take care that the different functional logics retain their own right (Wiarda, 2015, pp. 286f.) Thus, the problem, again, is to find the proper balance. It is difficult to make a forecast here whether this balancing process will succeed. One can probably expect that there will be times when the balancing works well and times when the balancing works less well and, in cases when the MoD's enlightened self-interest becomes weaker.

In general, it can be stated that military sociology is firmly established and effective as a policy-advising sociology. As a public sociology, it even influences social debates sometimes. To intensify this, military sociology at the ZMSBw not only has to work properly and with scientific excellence, but perhaps needs to establish stronger links to society and the public in the sense of a public sociology. The remaining dissatisfaction of many military sociologists with their own subject is mainly due to being ignored widely in professional sociology (Biehl, 2020). Yet this is a problem inside the academic world.

Notes

1 Translations from German into English language have been done by the authors.

2 The Namibia-born German political scientist Dieter Mahncke was a wanderer between the worlds of science and politics. After his studies at the University of Cape Town, the University of North Carolina at Chapel Hill and at Johns

Hopkins University in Baltimore, he received his doctorate in 1968 at Johns Hopkins' School of Advanced International Studies. He then worked for five years in Bonn as a research assistant in the German Council on Foreign Relations (DGAP), which since its very beginnings, had close working relations to decision-makers in politics, the media, the economy, and the justice system. After his habilitation at the University of Bonn in 1974, he held a chair in Political Science at the University of the Bundeswehr in Hamburg until 1980. In 1980, Mahncke switched to politics and took over the management of the speech writing department of the Federal Presidency. Five years later, he became Deputy Head of Planning Staff in the German Department of Defense (DoD). In 1996, he returned to science and taught European foreign, security and defense policy at the College of Europe in Brussels and Warsaw until his retirement in 2010.

3 Perhaps even more illustrative is the case of dealing with homosexuality in the US military. Here, military sociologist Charles C. Moskos devised the Solomonian formula of "Don't ask, don't tell" in the early 1990s, thus enabling President Bill Clinton to renounce the strict rejection of homosexuals in the US military in his 1992 election campaign. Clinton himself referred to Moskos as the spiritus rector of this policy in American television several times (N.N., 2009). On the other hand, Moskos' proposal to reintroduce conscription was not taken up by the American administration.

4 That is why the research on the consulting business has become more intense since the mid-1990s (see, e.g., vgl. etwa Murswieck, 1994c; Alemann & Vogel, 1996; Howaldt & Kopp, 1998; Elbe, 2001; Degele et al., 2001; Weingart & Lentsch, 2008; Elbe & Saam 2008; see also Wissenschaftszentrum Berlin, 1977).

References

Adorno, Th. W., Dahrendorf, R., Pilot, H., Albert, H., Habermas, J., & Popper, K. R. (1972). *Der Positivismusstreit in der deutschen Soziologie*. Neuwied – Berlin: Luchterhand.

Alemann, H. von, & Vogel, A. (Eds.) (1996). *Soziologische Beratung: Praxisfelder und Perspektiven*. Opladen: Leske + Budrich.

Allmendinger, J. (Ed.) (2001). *Gute Gesellschaft? Verhandlungen des 30. Kongresses der Deutschen Gesellschaft für Soziologie in Köln 2000. Teil B*. Opladen: Leske + Budrich.

Badura, B. (1980). Gegenexpertise als wissenschaftssoziologisches und wissenschaftspolitisches Problem. *Soziale Welt* 31: 459–473.

Beck, U. (1980). Die Vertreibung aus dem Elfenbeinturm. Anwendung soziologischen Wissens als soziale Konfliktsteuerung. *Soziale Welt* 31: 415–441.

Beck, U. (Ed.) (1982). *Soziologie und Praxis. Erfahrungen, Konflikte, Perspektiven*. Göttingen: Otto Schwartz Verlag.

Beck, U., & Bonß, W. (1984). Soziologie und Modernisierung. Zur Ortsbestimmung der Verwendungsforschung. *Soziale Welt* 35: 381–406.

Beck, U., & Bonß, W. (1989). Verwissenschaftlichung ohne Aufklärung? Zum Strukturwandel von Sozialwissenschaft und Praxis. In U. Beck & W. Bonß (Eds.), *Weder Sozialtechnologie noch Aufklärung? Analysen zur Verwendung sozialwissenschaftlichen Wissens* (pp. 7–45). Frankfurt am Main: Suhrkamp.

Becker, H., & Hager, F. (1992). *Aufklärung als Beruf*. Munich: Piper.

Biehl, H. (2020). Funktionen militärsoziologischer Forschung. Eine Bilanz der Militärsoziologie als empirische Sozialforschung. In M. Elbe, H. Biehl &

M. Steinbrecher (Eds.), *Empirische Sozialforschung in den Streitkräften. Positionen, Erfahrungen, Kontroversen*. Berlin: Berliner Wissenschaftsverlag. [forthcoming].

Biehl, H., & Keller, J. (2016). Ein anderer Blick auf den Einsatz. Die Forschung des SOWI zu Auslandseinsätzen. In A. Dörfler-Dierken & G. Kümmel (Eds.), *Am Puls der Bundeswehr. Militärsoziologie in Deutschland zwischen Wissenschaft, Politik, Bundeswehr und Gesellschaft* (pp. 189–205). Wiesbaden: Springer VS.

Bolz, N. (2001). Es geht auch anders. Ethik als Versteck und die Grenzen der Politikberatung. *Frankfurter Rundschau*, 29 May 2001, 19.

Bonacker, Th. (2016). Zwischen wissenschaftlicher Autonomie und politischen Vorgaben: Die Evaluation der Ressortforschung. In A. Dörfler-Dierken & G. Kümmel (Eds.), *Am Puls der Bundeswehr. Militärsoziologie in Deutschland zwischen Wissenschaft, Politik, Bundeswehr und Gesellschaft* (pp. 105–121). Wiesbaden: Springer VS.

Bonß, W., & Hartmann, H. (Eds.) (1985). *Entzauberte Wissenschaft. Zur Relativität und Geltung soziologischer Forschung*. Göttingen: Schwarz.

Borck, C., Lipphardt, V., Maasen, S., & Müller, R. (2018). Responsible Research? Dilemmata der Integration gesellschaftlicher und kultureller Perspektiven in naturwissenschaftliche Forschungsprogramme. *Beiträge zur Wissenschaftsgeschichte* 41: 215–221.

Bredow, W. von (1999). Ein verläßlicher wissenschaftlicher Stützpunkt. Universitäre Lehre und Forschung über das Militär wären in den Sozialwissenschaften in Deutschland ohne das SOWI schon lange ausgetrocknet. In Sozialwissenschaftliches Institut der Bundeswehr (Ed.), Sozialwissenschaftliches Institut der Bundeswehr (Ed.), *25 Jahre Sozialwissenschaftliches Institut der Bundeswehr. Eine zukunftsorientierte Standortbestimmung* (pp. 45–52). Strausberg: SOWI.

Breitbach, B. (1987). *Bericht über Bruno*. Frankfurt am Main: Fischer.

Brodocz, A., & Noetzel, Th. (2004). Auf dem Weg zu einer konstruktivistischen Politikberatung. In G. Kümmel (Ed.), *Wissenschaft, Politik und Politikberatung. Erkundungen zu einem schwierigen Verhältnis* (pp. 193–214). Frankfurt am Main et al.: Peter Lang.

Brückweh, K., Schumann, D., Wetzell, R., & Ziemann, B. (Eds.) (2012). *Engineering society. The role of the human and social sciences in modern societies, 1880–1980*. Basingstoke: Palgrave Macmillan.

Buchholz, K. (2008). *Professionalisierung der wissenschaftlichen Politikberatung? Interaktions- und professionssoziologische Perspektiven*. Bielefeld: Transcript.

Bundesministerin der Verteidigung Ursula von der Leyen (2015). *Rede in der Debatte über den Verteidigungshaushalt. Plenarprotokoll 18/139, Stenografischer Bericht, 139. Sitzung, 25.11.2015* (pp. 13670–13672). Berlin: Deutscher Bundestag. Retrieved from http://dipbt.bundestag.de/dip21/btp/18/18139.pdf#P.13669; 27 January 2019.

Bundesministerium der Verteidigung (BMVg) (1984). *Statut für das Sozialwissenschaftliche Institut der Bundeswehr (Neufassung) vom 28. September 1984*. Bonn: BMVg.

Bundesministerium der Verteidigung (BMVg) (1997). *Aufgabenkonzeption für das Sozialwissenschaftliche Institut der Bundeswehr (SWInstBw), erlassen vom Stellvertreter des Generalinspekteurs der Bundeswehr am 26. February 1997*. Bonn: BMVg.

Bundesministerium der Verteidigung (BMVg) (2007). *Statut für das Sozialwissenschaftliche Institut der Bundeswehr (SWInstBw) vom 19. April 2007*. Bonn: BMVg.

Bundesministerium der Verteidigung (BMVg), Führungsstab der Streitkräfte I 15 (1976). *Sozialwissenschaftliches Institut der Bundeswehr, Info 1.* Bonn: BMVg.

Burawoy, M. (2005). For public sociology. American Sociological Association Presidential Address 2004. *American Sociological Review* 70: 4–28.

Burth, H.-P., & Görlitz, A. (Eds.) (2001). *Politische Steuerung in Theorie und Praxis.* Baden-Baden: Nomos.

Chahoud, A. (2010). Blicke hinter die Kulissen: Aus dem 'Alltag' eines Politikberaters. *Aus Politik und Zeitgeschichte* B-19: 34–38.

Dahrendorf, R. (2001). Über die Machbarkeit der guten Ordnung. In J. Allmendinger (Ed.), *Gute Gesellschaft? Verhandlungen des 30. Kongresses der Deutschen Gesellschaft für Soziologie in Köln 2000. Teil B* (pp. 1330–1337). Opladen: Leske + Budrich.

Degele, N., Münch, T., Pongratz, H. J., & Saam, N. J. (Eds.) (2001). *Soziologische Beratungsforschung. Perspektiven für Theorie und Praxis der Organisationsberatung.* Opladen: Leske + Budrich.

Dörfler-Dierken, A., & Kümmel, G. (Eds.) (2016). *Am Puls der Bundeswehr. Militärsoziologie in Deutschland zwischen Wissenschaft, Politik, Bundeswehr und Gesellschaft.* Wiesbaden: Springer VS.

Dreier, H. (2010). *Wissenschaft und Politik.* Stuttgart: Steiner.

Elbe, M. (2001). Organisationsberatung: Kritik und Perspektiven aus soziologisch-verstehender Sicht. In H. Wüthrich, H. W. Winter & A. Philipp (Eds.), *Grenzen ökonomischen Denkens. Auf den Spuren einer dominanten Logik* (pp. 551–580). Wiesbaden: Gabler.

Elbe, M., & Saam, N. J. (2008). "Mönche aus Wien, bitte lüftets eure Geheimnisse." Über die Abweichung der Beratungspraxis von den Idealtypen der Organisationsberatung. *Gruppendynamik und Organisationsberatung. Zeitschrift für angewandte Sozialpsychologie* 39: 326–350.

Engel, F., Nestmann, F., & Sickendiek, U. (2018). Beratung: Alte Selbstverständnisse und neue Entwicklungen. In S. Rietmann & M. Sawatzki (Eds.), *Zukunft der Beratung. Von der Verhaltens- zur Verhältnisorientierung?* (pp. 83–115). Wiesbaden: Springer VS.

Fisch, S., & Rudloff, W. (Eds.) *Experten und Politik. Wissenschaftliche Politikberatung in geschichtlicher Perspektive.* Berlin: Duncker & Humblot.

Frank, H. (Vizeadmiral, stellvertretender Generalinspekteur der Bundeswehr) (1998). *Ansprache anläßlich der Übergabe des Sozialwissenschaftlichen Instituts der Bundeswehr von Oberst i.G. Geppert an Oberstleutnant i.G. Lohmann am 16. Dezember 1998 in Strausberg (Vortragsmanuskript).* Strausberg: SOWI.

Friedrichs, J., Lepsius, M. R., & Mayer, K. U. (1998). Diagnose und Prognose in der Soziologie. In J. Friedrich, M. R. Lepsius & K. U. Mayer (Eds.), *Die Diagnosefähigkeit der Soziologie* (pp. 9–31). Opladen: Westdeutscher Verlag.

Froese, A., Mevissen, N., Böttcher, J., Simon, D., Lentz, S., & Knie, A. (2014). *Wissenschaftliche Güte und gesellschaftliche Relevanz der Sozial- und Raumwissenschaften: Ein spannungsreiches Verhältnis. Handreichung für Wissenschaft, Wissenschaftspolitik und Praxis.* Berlin: Wissenschaftszentrum Berlin für Sozialforschung (WZB).

Gellner, W. (1994): Politikberatung durch nichtstaatliche Akteure – Typen, Funktionen, Strategien. In A. Murswieck (Ed.), *Regieren und Politikberatung* (pp. 175–192). Opladen: Leske + Budrich.

Geppert, H. (1998). *Die Militärsoziologie aus der Sicht des Bedarfsträgers*. Strausberg: SOWI.

Greven, M. Th. (1999). *Die politische Gesellschaft. Kontingenz und Dezision als Probleme des Regierens und der Demokratie*. Opladen: Leske + Budrich.

Habermas, J. (1973): Wahrheitstheorien. In H. Fahrenbach (Ed.), *Wirklichkeit und Reflexion. Walter Schulz zum 60. Geburtstag* (pp. 211–265). Pfullingen: Verlag Günther Neske.

Habermas, J., & Luhmann, N. (1985). *Theorie der Gesellschaft oder Sozialtechnologie – Was leistet die Systemforschung?* Frankfurt am Main: Suhrkamp.

Hagen, U. vom (2012). *Homo militaris. Perspektiven einer kritischen Militärsoziologie*. Bielefeld: Transcript.

Hauer, I. (1999). Ein Institut in der Krise. *Sicherheit & Frieden* 17: 75–76.

Heinemann, W. (2016). Das SOWI im Lichte der Akten. In A. Dörfler-Dierken & G. Kümmel (Eds.), *Am Puls der Bundeswehr. Militärsoziologie in Deutschland zwischen Wissenschaft, Politik, Bundeswehr und Gesellschaft*. (pp. 35–50). Wiesbaden: Springer VS.

Honneth, A. (2001). Wie ist eine gute Gesellschaft denkbar und gestaltbar? Thesen zu einer unübersichtlichen Diskussionslandschaft. In J. Allmendinger (Ed.), *Gute Gesellschaft? Verhandlungen des 30. Kongresses der Deutschen Gesellschaft für Soziologie in Köln 2000. Teil B* (pp. 1322–1329). Opladen: Leske + Budrich.

Hoppe, R. (2005). Rethinking the science-policy nexus: From knowledge utilization and science technology studies to types of boundary arrangements. *Poiesis & Praxis* 3: 199–215.

Howaldt, J. (2008). Die Soziologie in Zeiten der Wissensgesellschaft – Kritische Anmerkungen zu einer unzeitgemäßen Unterscheidung. In D. Euler, J. Howaldt, G. Reinmann & R. Weiß (Eds.), *Neue Forschungsverständnisse in den Sozialwissenschaften: Konsequenzen für die Berufsbildungsforschung im Bundesinstitut für Berufsbildung* (pp. 27–42). Bonn: Bundesinstitut für Berufsbildung.

Howaldt, J., & Kopp, R. (Eds.) (1998). *Sozialwissenschaftliche Organisationsberatung. Auf der Suche nach einem spezifischen Beratungsverständnis*. Berlin: Edition Sigma.

Hustedt, T., Veit, S., & Fleischer, J. (2010). Wissen ist Macht? Wissenschaftliche Politikberatung. *Aus Politik und Zeitgeschichte* B-19: 15–21.

Jann, W. (1994). Wissenschaftler in der Regierung – Advokatoren der Verwissenschaftlichung der Politik? In A. Murswieck (Ed.), *Regieren und Politikberatung* (pp. 159–173). Opladen: Leske + Budrich.

Jung, K. (Parlamentarischer Staatssekretär) (1983). *Ansprache des Herrn Parlamentarischen Staatssekretärs Dipl.-Ing. Jung anläßlich des Ausscheidens von Professor Dr. Ralf Zoll aus dem Amt des Direktors des SOWI der Bundeswehr in München am 31. Januar 1983*. Munich: SOWI.

Kaufmann, F.-X. (1976). Zum Verhältnis von Soziologie und Politik – Das Beispiel Zweiter Familienbericht. *Zeitschrift für Soziologie* 5: 285–306.

Klein, P. (2004). Das Sozialwissenschaftliche Institut der Bundeswehr in der Politikberatung. In G. Kümmel (Ed.), *Wissenschaft, Politik und Politikberatung. Erkundungen zu einem schwierigen Verhältnis* (pp. 31–49). Frankfurt am Main et al.: Peter Lang.

König, R. (1968). Einige Bemerkungen zu den speziellen Problemen der Begründung einer Militärsoziologie. In R. König (Ed.), Beiträge zur Militärsoziologie (pp. 7–12). Köln – Opladen: Westdeutscher Verlag.

Kühl, S. (2003). Wie verwendet man Wissen, das sich gegen die Verwendung sträubt? Eine professionssoziologische Neubetrachtung der Theorie-Praxis-Diskussion in der Soziologie. In H.-W. Franz, J. Howaldt, H. Jacobsen & R. Kopp (Eds.), *Forschen – lernen – beraten. Der Wandel von Wissensproduktion und -transfer in den Sozialwissenschaften* (pp. 71–92). Berlin: Edition Sigma.

Kümmel, G. (2016). Halb zog man sie, halb sank sie hin … Die Bundeswehr und ihre Öffnung für Frauen. In A. Dörfler-Dierken & G. Kümmel (Eds.), *Am Puls der Bundeswehr. Militärsoziologie in Deutschland zwischen Wissenschaft, Politik, Bundeswehr und Gesellschaft* (pp. 277–301). Wiesbaden: Springer VS.

Leif, Th. (2006). *Beraten und verkauft: McKinsey & Co. – der große Bluff der Unternehmensberater.* 10th Edition. Munich: C. Bertelsmann Verlag

Lippert, E., & Wachtler, G. (1982). Militärsoziologie – eine Soziologie ‚nur für den Dienstgebrauch'? In U. Beck (Ed.), *Soziologie und Praxis. Erfahrungen, Konflikte, Perspektiven* (pp. 335–355). Göttingen: Otto Schwartz Verlag.

Luhmann, N. (1977). Theoretische und praktische Probleme der anwendungsbezogenen Sozialwissenschaften: Zur Einführung. In Wissenschaftszentrum Berlin (Ed.), *Interaktion von Wissenschaft und Politik. Theoretische und praktische Probleme der anwendungsorientierten Sozialwissenschaften* (pp. 16–39). Frankfurt am Main – New York: Campus.

Luhmann, N. (1989). *Vertrauen. Ein Mechanismus der Reduktion sozialer Komplexität.* 3rd revised edition. Stuttgart: Ferdinand Enke.

Luhmann, N. (1991). *Die Wissenschaft der Gesellschaft.* Frankfurt am Main: Suhrkamp.

Luhmann, N. (2005). Theoretische und praktische Probleme der anwendungsbezogenen Sozialwissenschaften. In N. Luhmann (Ed.), *Soziologische Aufklärung 3.* 4th edition (pp. 369–385). Wiesbaden: Westdeutscher Verlag.

Maasen, S., Elberfeld, J., Eitler, P., & Tändler, M. (Eds.) (2011). *Das beratene Selbst: Zur Genealogie der Therapeutisierung in den 'langen' Siebzigern.* Bielefeld: Transcript.

Mackewitsch, R. (1999). Sozialwissenschaftliche Forschung als Politikberatung. In Sozialwissenschaftliches Institut der Bundeswehr (Ed.), *25 Jahre Sozialwissenschaftliches Institut der Bundeswehr. Eine zukunftsorientierte Standortbestimmung* (pp. 75–95). Strausberg: SOWI.

Mahncke, D. (1994). *The role of civilian experts within the defence community. Paper presented at the 1994 International Research Seminar on Euro-Atlantic Security, Democratic and Civil Control over Military Forces.* Rom: NATO Defense College.

Mayntz, R. (1994). Politikberatung und politische Entscheidungsstrukturen: Zu den Voraussetzungen des Politikberatungsmodells. In A. Murswieck (Ed.), *Regieren und Politikberatung* (pp. 17–29). Opladen: Leske + Budrich.

Mielke, G. (1999). Die politische Planung steckt in der Zwickmühle. Über die Struktur und Qualität sozialwissenschaftlicher Beratung in den Staatskanzleien der Länder. Ein Werkstattbericht. *Frankfurter Rundschau*, 27 July 1999, 18.

Mulisch, H. (1995 [1992]). *Die Entdeckung des Himmels.* Reinbek bei Hamburg: Rowohlt.

Murswieck, A. (1994a): Einleitung. In A. Murswieck (Ed.), *Regieren und Politikberatung* (pp. 9–13). Opladen: Leske + Budrich.

Murswieck, A. (1994b): Wissenschaftliche Beratung im Regierungsprozeß. In A. Murswieck (Ed.), *Regieren und Politikberatung* (pp. 103–119). Opladen: Leske + Budrich.

Murswieck, A. (Ed.) (1994c). *Regieren und Politikberatung*. Opladen: Leske + Budrich.

N.N. (2009). *Charles C. Moskos*. Retrieved from http://charlesmoskos.com/index.htm; 27 January 2019.

Neun, O. (2016). Die Verwendungsdebatte innerhalb der deutschen Soziologie: Eine vergessene Phase der fachlichen Selbstreflexion. In H. Staubmann (Ed.), *Soziologie in Österreich – Internationale Verflechtungen* (pp. 333–353). Innsbruck: Innsbruck University Press.

Nullmeier, F. (2011). Max Weber Revisited: Zum Verhältnis von Politik und Wissenschaft. Oder: Politikberatung als Nebenberuf. In W. Lamping & H. Schridde (Eds.), *Der konsultative Staat. Reformpolitik und Politikberatung* (pp. 39–46). Opladen: Verlag Barbara Budrich.

Offe, C. (1977). Die kritische Funktion der Sozialwissenschaften. In Wissenschaftszentrum Berlin (Ed.), *Interaktion von Wissenschaft und Politik. Theoretische und praktische Probleme der anwendungsorientierten Sozialwissenschaften* (pp. 321–329). Frankfurt am Main – New York: Campus.

Personalstrukturkommission des Bundesministers der Verteidigung (1971). *Die Personalstruktur der Streitkräfte*. Bonn: BMVg.

Pielke, R. A., Jr. (2007). *The honest broker. Making sense of science in policy and politics*. Cambridge: Cambridge University Press.

Presse- und Informationsamt der Bundesregierung (1971). *Weißbuch 1971/72. Zur Sicherheit der Bundesrepublik Deutschland und zur Entwicklung der Bundeswehr*. Bonn: Deutscher Bundestag.

Prisching, M. (2000). Bilder von Wissenschaftlern in Literatur und Realität. In O. Panagl, & W. Weiss (Eds.) (2000). *Noch einmal Dichtung und Politik. Vom Text zum politisch-sozialen Kontext, und zurück* (pp. 31–84). Vienna – Cologne – Graz: Böhlau Verlag.

Raphael, L. (2018). Die Verwissenschaftlichung des Sozialen – Wissens- und Sozialordnungen im Europa des 20. Jahrhunderts. In L. Raphael, *Ordnungsmuster und Deutungskämpfe. Wissenspraktiken im Europa des 20. Jahrhunderts* (pp. 13–49). Göttingen: Vandenhoeck & Ruprecht.

Rehberg, K.-S. (2001). Eine Wiederbegegnung von Politik und Soziologie? Einführung in die Abendveranstaltung. In J. Allmendinger (Ed.), *Gute Gesellschaft? Verhandlungen des 30. Kongresses der Deutschen Gesellschaft für Soziologie in Köln 2000. Teil B* (pp. 1317–1321). Opladen: Leske + Budrich.

Richter, G. (Eds.) (2012). *Neuausrichtung der Bundeswehr. Beiträge zur professionellen Führung und Steuerung*. Wiesbaden: Springer VS.

Rietmann, S., & Sawatzki, M. (Eds.) (2018). *Zukunft der Beratung. Von der Verhaltens- zur Verhältnisorientierung?* Wiesbaden: Springer VS.

Ronge, V. (1989). Verwendung sozialwissenschaftlicher Ergebnisse in institutionellen Kontexten. In U. Beck & W. Bonß (Eds.), *Weder Sozialtechnologie noch Aufklärung? Analysen zur Verwendung sozialwissenschaftlichen Wissens* (pp. 332–354). Frankfurt am Main: Suhrkamp.

Rüb, F. W. (2011). Inkompatible Welten? Politik und die Grenzen wissenschaftlicher Politikberatung. In W. Lamping & H. Schridde (Eds.), *Der konsultative Staat. Reformpolitik und Politikberatung* (pp. 63–87). Opladen: Verlag Barbara Budrich.

Saam, N. J., & Petran, W. (2001). Einleitung. In N. Degele, T. Münch, H. J. Pongratz, & N. J. Saam (Eds.), *Soziologische Beratungsforschung. Perspektiven für Theorie und Praxis der Organisationsberatung* (pp. 7–13). Opladen: Leske + Budrich.

Sala, R. (2017). Verwissenschaftlichung des Sozialen – Politisierung der Wissenschaft? Zum Verhältnis von Wissenschaft und Politik in der Geschichtsschreibung des 19. und 20. Jahrhunderts. *Berichte zur Wissenschaftsgeschichte* 40: 333–349.

Schelsky, H. (1961). Der Mensch in der wissenschaftlichen Zivilisation. In H. Schelsky, *Auf der Suche nach Wirklichkeit. Gesammelte Aufsätze* (pp. 439–480). Düsseldorf – Cologne: Eugen Diederichs Verlag.

Schluchter, W. (1971). *Wertfreiheit und Verantwortungsethik: Zum Verhältnis von Wissenschaft und Politik bei Max Weber*. Tübingen: Mohr.

Schneider, W. L. (1989). Kooperation als strategischer Prozeß. Administrative Auftragsforschung im Spannungsfeld zwischen professionellem Interesse und politischer Instrumentalisierung. In U. Beck & W. Bonß (Eds.), *Weder Sozialtechnologie noch Aufklärung? Analysen zur Verwendung sozialwissenschaftlichen Wissens* (pp. 302–331). Frankfurt am Main: Suhrkamp.

Schneidewind, U., & Singer-Brodowski, M. (Eds.) (2014). *Transformative Wissenschaft. Klimawandel im deutschen Wissenschafts- und Hochschulsystem*. 2nd edition. Marburg: Metropolis.

Schubert, K. von (1983). *Ansprache anläßlich des Ausscheidens von Professor Dr. Ralf Zoll aus dem Amt des Direktors des SOWI der Bundeswehr in München am 31. January 1983*. Munich: SOWI.

Schützeichel, R. (2008). Beratung, Politikberatung, wissenschaftliche Politikberatung. In S. Bröchler & R. Schützeichel (Eds.), *Politikberatung* (pp. 5–23). Stuttgart: Lucius & Lucius.

Schützeichel, R., & Brüsemeister, Th. (2004a): Einleitung. In R. Schützeichel, & Th. Brüsemeister (Eds.), *Die beratene Gesellschaft. Zur gesellschaftlichen Bedeutung von Beratung* (pp. 7–18). Wiesbaden: VS Verlag für Sozialwissenschaften.

Schützeichel, R., & Brüsemeister, Th. (Eds.) (2004b). *Die beratene Gesellschaft. Zur gesellschaftlichen Bedeutung von Beratung*. Wiesbaden: VS Verlag für Sozialwissenschaften.

Sohst, J. (Brigadegeneral, Stabsabteilungsleiter im Führungsstab der Streitkräfte) (1999). Zielsetzungen sozialwissenschaftlicher Forschung in der Bundeswehr, Aufgabenwahrnehmung durch das Sozialwissenschaftliche Institut und künftige Arbeitsschwerpunkte aus der Sicht des Auftraggebers. In Sozialwissenschaftliches Institut der Bundeswehr (Ed.), *25 Jahre Sozialwissenschaftliches Institut der Bundeswehr. Eine zukunftsorientierte Standortbestimmung* (pp. 71–74). Strausberg: SOWI.

Sozialwissenschaftliches Institut der Bundeswehr (Ed.) (2009): *35 Jahre Sozialwissenschaftliches Institut der Bundeswehr. Mehr verstehen. Besser entscheiden*. Strausberg: SOWI.

Sozialwissenschaftliches Institut der Bundeswehr (2012). *Newsletter des Sozialwissenschaftlichen Instituts der Bundeswehr. sowi.news 2/2012*. Strausberg: SOWI.

Spaemann, R. (1990). Ars longa vita brevis. In R. Löw, R. Spaemann, & P. Koslowski (Eds.), *Expertenwissen und Politik* (pp. 15–26). Weinheim: VCH, Acta Humaniora.

Stehr, N. (2001). Moderne Wissensgesellschaften. *Aus Politik und Zeitgeschichte* B 36: 7–14.

Stengers, I. (1998). *Wem dient die Wissenschaft?* Munich: Gerling Akademie Verlag.

Streeck, W. (2009). *Man weiß es nicht genau: Vom Nutzen der Sozialwissenschaften für die Politik.* Cologne: Max-Planck-Institut für Gesellschaftsforschung.

Strohschneider, P. (2014). Zur Politik der Transformativen Wissenschaft. In A. Brodocz, D. Herrmann, R. Schmidt, D. Schulz & J. Schulze Wessel (Eds.), *Die Verfassung des Politischen. Festschrift für Hans Vorländer* (pp. 175–192). Wiesbaden: Springer VS.

Szöllösi-Janze, M. (2004). Politisierung der Wissenschaften – Verwissenschaftlichung der Politik. Wissenschaftliche Politikberatung zwischen Kaiserreich und Nationalsozialismus. In S. Fisch & W. Rudloff (Eds.), *Experten und Politik. Wissenschaftliche Politikberatung in geschichtlicher Perspektive* (pp. 79–100). Berlin: Duncker & Humblot.

Weber, M. (1917/19). Wissenschaft als Beruf. In W. J. Mommsen, W. Schluchter & B. Morgenbrod (Eds.) (1994), *Max Weber Wissenschaft als Beruf 1917/1919. Politik als Beruf 1919. Studienausgabe der Max-Weber-Gesamtausgabe Vol. I/17* (pp. 1–23). Tübingen: J. C. B. Mohr (Paul Siebeck).

Weingart, P. (2003). *Wissenschaftssoziologie*. Bielefeld: Transcript.

Weingart, P., & J. Lentsch (2008). *Wissen, Beraten, Entscheiden: Form und Funktion wissenschaftlicher Politikberatung in Deutschland.* Weilerswist: Velbrück.

Wiarda, J.-M. (2015). *Was macht die Beratung mit dem Berater? Über die Folgen von Politikberatung für die Wissenschaft am Beispiel des Max-Planck-Instituts für Bildungsforschung*. Diss. phil. Kultur-, Sozial- und Bildungswissenschaftliche Fakultät. Berlin: Humboldt University.

Wissenschaftsrat (2000). *Thesen zur zukünftigen Entwicklung des Wissenschaftssystems in Deutschland (Drs. 4594/00)*. Cologne: Wissenschaftsrat.

Wissenschaftsrat (2009). *Wissenschaftspolitische Stellungnahme zum Sozialwissenschaftlichen Institut der Bundeswehr (SWInstBw), Strausberg (Drs. 9500-09)*. Aachen: Wissenschaftsrat.

Wissenschaftszentrum Berlin (Ed.) (1977). *Interaktion von Wissenschaft und Politik. Theoretische und praktische Probleme der anwendungsorientierten Sozialwissenschaften.* Frankfurt am Main – New York: Campus.

Zentrum für Militärgeschichte und Sozialwissenschaften der Bundeswehr (ZMSBw), Kommandeur (2017). *Geschäftsordnung des Zentrums für Militärgeschichte und Sozialwissenschaften der Bundeswehr (GO ZMSBw)*. Potsdam: ZMSBw.

Zoll, R. (1999): Back to the Roots! In Sozialwissenschaftliches Institut der Bundeswehr (Ed.), *25 Jahre Sozialwissenschaftliches Institut der Bundeswehr. Eine zukunftsorientierte Standortbestimmung* (pp. 123–132). Strausberg: SOWI.

Zoll, R. (2016). Von wissenschaftlicher Politikberatung zur militärischen Dienststelle. Eine Kurzgeschichte des Sozialwissenschaftlichen Instituts der Bundeswehr (SOWI). In A. Dörfler-Dierken & G. Kümmel (Eds.), *Am Puls der Bundeswehr. Militärsoziologie in Deutschland zwischen Wissenschaft, Politik, Bundeswehr und Gesellschaft* (pp. 21–33). Wiesbaden: Springer VS.

3 What does social science research on the military "do"?

Between Academic Ornament and Dubious Ally of the Israeli Military

Eyal Ben-Ari

This chapter offers a personal reflection about over three and half decades of research on the Israeli armed forces. The perspective I take looks at the dynamics by which my research-related participation in various academic, public, and military arenas led to some (rather limited) changes in the Israel Defence Forces (IDF). Accordingly, I trace out the social encounters through which my research (often in cooperation with academic colleagues and serving officers) was amalgamated into forms that military actors used to alter some policies, structures, practices, or ideas. In what follows, I suggest that the (again, rather limited) impact of my work should be understood in terms of the formal and informal (or hidden) agendas of actors (including myself) and the unintended (and usually unforeseen) implications of my very participation (see introduction, this volume).

Further, with regard to my own case, I pose a wider question, namely about what does academic participation in creating knowledge about and for the military "do"? This kind of question demands situating ourselves within the various organizational, institutional, and social networks and coalitions centering on the military. It implies understanding that researching the armed forces – expressed in such issues as gaining access and trust, gathering data, interpreting findings, or publishing – are all essentially social behaviors through which the "impact" we have is expressed. As I will show, my work, like that of many other scholars, served variously to "ornament" discussions, legitimate policy, lend scientific credibility to arguments, support allies in the military, change practices, open up new areas for investigation, set standards for academic research, or contribute to public debates.

A short note on a long personal journey

To the best of my knowledge, I am probably "the" Israeli anthropologist who is most identified with the study of the armed forces (Ben-Ari, forthcoming). After serving four years in the 1970s as a conscript and then as an officer, I then completed 22 years in the reserves, ending at the rank of major. The seeds of my work in the military were sown during the war of 1973 when I was

DOI: 10.4324/9781003165217-3

an infantry soldier and wounded by an Egyptian sniper in the city of Suez. It was during that war that I began (at that time only vaguely) to be aware of questions about Israel's military might (and vulnerability) and my own political commitments. Upon being discharged from the IDF, I went on to study sociology and anthropology at the Hebrew University of Jerusalem. During those years, I served as an officer in a reserve infantry battalion usually undertaking about 30–40 days of service each year.

In 1980 I decided to pursue my PhD in social anthropology and completed my studies at Cambridge University in 1984 with a thesis about Japanese white-collar communities (Ben-Ari, 1991). Upon returning to Israel, I joined another infantry battalion as an adjutant and was finally discharged from the IDF at the age of forty-four. Having honed my anthropological skills on Japanese material (I continue to study Japan), the direct origin of my scholarly interest in the armed forces lies in the deployment of my battalion to the Occupied Palestinian Territories before and during the first Palestinian Uprising (1988–1991). To make sense of my experience as a troubled citizen during that time, I used my ethnographic training to research and then publish an article about the battalion I served in (Ben-Ari, 1989). As I wrote this piece, I felt that I needed to ask wider questions about the significance of military service, the transition between civilian and army lives, or how violence used by Israeli troops is explained and justified. Hence between 1987 and 1992 (when I left Israel for a sabbatical) I undertook a systematic study of the battalion. This effort culminated in a volume that can be playfully characterized as a form of "grunt ethnography" (Ben-Ari, 1998). The ethnography covered the ethno-models troops used to make sense of their environment and wider questions about manhood and motivation related to the Israeli military.

In the mid-1990s I decide to add the sociology of the military to my research fields of Japanese communities and Japanese early childhood education. From that period onward, I began to conduct projects related to "things military" that I carried out alone or much more often with other scholars. In addition, I began to supervise masters and PhD students studying such issues. Eventually, 24 of my masters and 13 of my doctoral students worked on various studies of the military and I found myself frequently co-supervising them with colleagues from sociology, anthropology, social psychology, history, political science, or organization studies. Finding subjects to study in Israel is relatively easy because Israel is involved in a militarized occupation since 1967, has participated in major wars and operations since its inception, and sees sporadic but constant armed attacks. In addition, being a relatively open society, it is quite easy to obtain interviews and (with varying degrees of difficulty) access to the IDF. These conditions meant that I had a rather 'rich' source of material to undertake research for more than 35 years. In addition, since I know Japanese, since 1998 I have also been studying the Japanese Self-Defense Forces and, more recently, the Taiwanese military. In this chapter, I limit myself to the IDF.

Some key activities

I present five types of activities I participated in in order to illuminate the diversity of my "impacts" and the different kinds of formal and informal aims, the direct and indirect uses of the knowledge produced, and the projects' academic products.

Combat companies in the second Palestinian Uprising

In the spring of 2000, I was approached by the Behavioral Sciences Section of the IDF ground forces to carry out research on combat companies. The project was carried out with three serving behavioral science officers: a sociologist, a social-psychologist, and an anthropologist (Ben-Ari et al., 2009). In the beginning, the formal aim was to analyze the social dynamics and leadership patterns of these units. In fact, this was a period when Israel had withdrawn from southern Lebanon and when, we assumed, the main operational assignments of the IDF would continue to be policing the Occupied Territories and implementing the Oslo Accords. Three short months after we began, the second Palestinian Uprising, the Al-Aqsa Intifada, erupted and marked the IDF's sudden transition into a prolonged conflict. We consequently found ourselves in a rather "advantageous" situation of being able to chart-out this transition and accompany the development of the conflict. For us social scientists, the Al-Aqsa Intifada provided a rare opportunity to witness how military forces shift quickly from routine activities into sudden, often violent, action. Hence the formal aim of the project was adapted to the new circumstances of the uprising.

Apart from the project's formal aim, my aim emanated from my research interests in combat soldiers and the opportunity to study local-level units in action. The volume (and a series of articles) that emerged from this project that ended in 2006 (Ben-Ari, 2011) analyzed the creation of ad hoc temporary units, armed fighting in urban areas, the role of women in supporting the combat companies as drivers and spotters along the country's border, and the combination of violent practices and limits characterizing the Israeli armed forces' actions. Funding for this project was provided by the Israeli military and used for hiring research assistants (my students). Some of these students, in turn, wrote their masters' theses as part of this project.

Because this project involved me in a cooperative venture with serving officers, we found ourselves formulating research questions in abstract terms such as "new forms of tactical organizations" or conceptual frameworks suitable to the "new wars". We did this to overcome potential opposition from the IDF Spokesperson functionaries who were charged with producing and maintaining the army's good image. In effect, however, the final text included a heavy emphasis on the problems of dealing with civilians, fighting insurgents, how global human rights have impacted local level troops, and the dire impact of Israeli policy on Palestinian lives.

To my mind, there was no way I could have gained access to the kind of data we gathered if I were solely dependent on interviewing soldiers outside the military context. Gaining access, in turn, is related to the potential impact of my kind of research. While I do not know of any direct effects of the research, my coresearchers who were military officers did appear in internal IDF forums to speak about how the ground forces used violence in the Territories. More widely, reports about this research and my lectures inside and outside the IDF were part of the public debate about the use of violence in this conflict. In all of these forums, the fact that the project was based on a very wide-ranging empirical base provided, I surmise, some scientific credibility to the arguments put forward. The indirect impact, however, was much clearer and involved creating social scientific knowledge within the IDF's Center for Behavioral Sciences that was later taught to successive cohorts of consultants and researchers and within the IDF Military Academy where parts of our findings worked their way into the curriculum. In addition, it was used in regard to studies of temporary teams in the military and work about jointness in the IDF. No less important, and this is a point I return to later, through this project and other activities I not only set academic standards for internal military research but no less important cultivated military allies (while they, in turn, cultivated me and my students outside the military).

Applied research: checkpoints

A study that emerged alongside the one on combat companies was about the numerous checkpoints, roadblocks, and barriers that the IDF used during the Second Intifada. In 2003, two army officers serving in the ground forces' reserve component approached me to help improve the military's performance at the checkpoints (arrayed around the Occupied Territories). They asked to meet me because there were many problems – such as abuse or neglect – that were developing in these sites due to processing huge numbers of people and to the conflicting orders the soldiers were given (Ben-Ari et al., 2004). For the vast majority of Palestinians, checkpoints were (and to a great extent, still are) the symbol of the military Occupation, of aggression, and of humiliation. For this reason, I agreed to cooperate with the army on the basis of a hope that I could ease the plight of Palestinians moving through checkpoints and subsequently secured funding through a large German Foundation. A few months after completing our report, I appeared before a military committee appointed to look into the checkpoints and met with a number of senior officers that commanded units in the West Bank.

The stated goal of the research was to understand and eventually improve the treatment given to people in the checkpoints while my aims were both to further academic knowledge about military occupations and to try and contribute to a more humane treatment of Palestinians by the IDF. The effect of this project can only be understood, I now understand, as one component (perhaps a minor one) in a much broader effort undertaken by the media and NGOs, human rights groups and lawyers, Palestinian opposition groups and academics, as well as the pressure applied by commanders within the IDF to deal with the checkpoints. The result, achieved over a number of years, was improvement in the infrastructure of the checkpoints, a drastic reduction in their number, turning some of them into permanent fixtures that were staffed by specially trained military police units or (along the Israel's borders) by civilian officials, and training programs for soldiers staffing these barriers. Our specific project thus entailed not only meeting senior officers and select military committees but also contributing to the public debate about the checkpoints (mainly through lectures and presentations in academic and semi-public forums).

I am aware that there may be critical readings of this project since instituting improvements at checkpoints can be seen as a sort of defense mechanism that grants myself and my military allies a self-image of being humane while allowing the military Occupation to continue. Thus, by making the checkpoints more compassionate, the argument would go, we were making the occupation take on a more "human" face thereby inadvertently legitimatizing its prolongation. While I accept these criticisms, I chose to contribute to what I see as the reduction of suffering even if in a localized and limited way (Ben-Ari, 2005, 2011). I decided to work within the system – to try and make some even limited changes – so as to create a difference to Palestinian lives, to reduce harm.

Pubic and (not-so public) committees

Since the mid-1990s, I have participated in tens of meetings and committees related to military issues. The meetings covered themes such as civil-military relations, gender relations in the IDF, managing diversity, legal restraints on the IDF in the Territories, cohesion, military leadership, or the place of orthodox religious soldiers in the forces. In the civilian sphere I have appeared in conferences organized by research institutes, social movements, and the national parliament. The in-house military activities took place in the IDF's behavioral sciences units, the IDF Leadership Development School, schools of the IDF Chief Educational Officer, brigade, division and territorial commands, select committees convened by generals or two committees organized separately by the Deputy and the Chief of Staff of the IDF. In all of these encounters, I was invited as an "expert", someone who had mastered the skills and knowledge of sociology and particularly of the sociology of the military. Most of these opportunities involved senior commanders wanting academic inputs in regard to policy decisions (through brain-storming) or educational occasions for soldiers and officers. They hence well fit the enlightenment model of military sociology that Moskos (1988) talks about.

Here let me mention two opportunities in detail since they illuminate very different dynamics. In 1998, I was invited to present my views at a workshop attended by about 15 officers headed by a general from the Chief Infantry and Paratrooper Division and devoted to examining "Company commanders in twenty first Century". I was the only civilian among the military officers and while their badges indicated their rank and role, my badge had my name and the designation of "Professor". As in a number of such encounters, my impression was that I was invited not so much for my analysis but rather as an academic ornament for the proceedings (much like the cases of Castro and Leirner in this volume). While I stated my views, they were typically ignored. Given that I quickly understood this point, I utilized these encounters to learn about how such IDF officers reasoned, about what they saw as important, and about the bureaucratic maneuvering they were undertaking. Reflecting about this meeting afterward, I further realized that participating as a scholarly "token" contributed to how the convening general situated himself within the career dynamics of the IDF. Thus, the minute impact that I did have was perhaps on the career of this officer and as a scholarly adornment to the meeting, as a means to award it a cachet of academic seriousness.

A more significant impact involved my participation in a group comprised of academics (including economists and the leading sociologist of the military in Israel, Yagil Levy) along with military organizational consultants that convened to analyze the state of Israel's military reserves and propose changes. I was invited to participate by a military ally (who headed the Ground Forces' behavioral sciences unit) and the IDF's Chief Reserve Officers (a general) whom I have known since high school. We met every two or three weeks (for about a year and a half), prepared material and proposals, and debated on them. The starting point for the whole process was a number of social movements headed by reservists and the disquiet of many commanders from the reserves about the state of that military component (Rein & Ben-Ari, forthcoming). Our committee was the first to link academic social science research to concrete proposals. Undergoing various permutations, our initial proposals were finally turned into law, one having an impact on the IDF through such measures as limiting the days of military service for reservists, forcing the IDF to report to parliament about the utilization of reservists, or offering material and non-material support for them (Levy, 2011). In addition, the fact that so many scholars participated in this committee conceivably helped to legitimate its work through our academic credentials.

The management of diversity in the IDF

In 2005, my friend and colleague Edna Lomksy-Feder and I were invited by the IDF's Center for Behavioral Sciences to take on a military-funded project on the management of diversity in the IDF. While we had already carried out a broad-based project that relied on previous academic work and media reports on this topic, this was an opportunity for direct access to the

IDF and for examining how different social and cultural groups were 'handled' at the local, unit level. Along with a team of very talented research assistants, we purposely chose five different kinds of units ranging from a combat battalion, to a technological unit, and on to a logistical base. The project resulted in a lengthy report for the IDF, a number of academic articles, and presentations both in university and military forums.

The formal aim of the project was to examine how diversity in the IDF was managed and emanated from changes in Israeli society and the influence of external thinking about forms of multiculturalism. However, the hidden aim of our internal military allies was learning about whether the IDF contributed to or changed social inequality. The gap between the explicit and implicit (hidden) agendas had to do with the staffing of some research positions in the IDF Behavioral Sciences Center by sociologists with a critical orientation. For the research to take place, however, it had to be formulated as contributing to the smooth running of the IDF. Indeed, because this cluster of issues is so politically sensitive, we had to phrase the research question as one marked but a functional rather than a critical edge. The direct impact of this project is very difficult to gauge since it is all but impossible to trace out its final effects on policy. Nevertheless, I can state with some certainty that it contributed, again, along with the activities of many other actors, to raising greater awareness of inequality and some limited measure of continuity in studying it.

Supervising students

The final kind of activity I describe involves supervision of military officers for advanced degrees. Out of the tens of students that I have supervised, around twenty PhDs and masters' theses were written by serving officers (I supervised other civilian students about the IDF as well). These students were primarily in the military's behavioral sciences units and ranged from the rank of major to full colonel (two even went on to become Heads of the IDF Center for Behavioral Sciences). Taken together, these supervisions were not only research projects but involved other impacts on the Israeli military. Let me explain.

For almost all of the students who were serving officers, advanced studies were a way to further their career both in material terms (graduate degrees are linked to higher salaries) and non-material terms (the prestige and the cultural capital of having such degrees) and by opening up employment opportunities upon ending their military service. For myself, I saw the goals of supervising students as developing further the ideas I had been developing about the armed forces, exploring new subjects of research, gaining access to the IDF, and as the sheer fun of working with people. The most direct impact of this supervisory work, I would suggest, is that of raising and setting the academic level of the IDF's social scientists. This impact was reinforced by lectures, workshops, and a number of a multi-year research group that met every month or so to exchange ideas about research on the IDF. This group comprised my students as well as students from all of Israel's universities as well as serving members of the military's behavioral sciences units.

Intertwined issues: secondary and tertiary impacts

As proposed in the introduction to this volume (Ben-Ari et al.), my work should be seen as part of a cluster of processes best captured by a "fuzzy" model of the impact of social science research-related activities on the armed forces. It is "fuzzy" because these impacts and the processes that bring them about are often unclear, ambiguous, and rather uncertain because as social scientists we are part of multiple entanglements between the civilian and military spheres (Sørensen & Ben-Ari, 2019). Within this model, military decision-makers (like all such leaders) are nodes around which actors are organized through nets of often loose alliances and counter-coalitions that act through organizational settings and social groups, and public debates and discussions. Further, still, these assorted coalitions are themselves not homogenous as, for example, within the armed forces – there are tensions between services, arms, and units (and individuals) and among academics. Conflict is very much part of the way knowledge is produced. Social scientists, then, are only one (frequently rather minor) part of such networks and alliances but *may* cumulatively, along with other actors, bring about change. For military leaders, moreover, reception of social scientific analyses involves various forms of acceptance, interpretation, adaptation, disregard or, indeed, even outright rejection. Against this background, let me explore a number of themes that are intertwined throughout my analysis.

With the exception of Yagil Levy, who regularly writes in the Haaretz newspaper (Israel's self-appointed beacon of progressive "thinking"), and of very occasional appearances of the heads of MAMDA in the media, military sociology is not very prominent in Israel's public forums (Ben-Ari, forthcoming). There is some activity of scholars in the new media through blogs and discussion forums but these are mostly limited in their breadth and often involve preaching to the converted. At the same time, much activity takes place in committees or internal reviews in which social science scholars participate. While not open to the general public, these forums nevertheless allow a form of participation in consultative processes. Finally, Israeli scholars studying the armed forces have regularly organized working groups, conferences, and workshops that are sometimes open to the public and which are occasionally reported by the media. I myself have been much more comfortable and relatively active in these latter two kinds of forums than in participating in Israel's public discussions.

I find that the impact of my own research activities has been achieved primarily through cooperative work with key allies within the IDF. These allies were very rarely commanders but rather members of the units belonging to the Center for the Behavioral Sciences of the IDF comprised of researchers, organizational consultants, and members of a unit charged with psychological evaluation of troops. These allies have staffed the units affiliated with the IDF Center for Behavioral Sciences and such as the offices of the Gender Advisor to the Chief of Staff, the Air Force, the IDF Leadership and Development School, the Home Front Command, or the Intelligence Corps (Ben-Ishai, 2017; Ben-Shalom/Fox, 2009; Lehrer/Amram-Katz, 2011; Sher/Ben-Eliyahu, 2011). These key allies were the ones that mediated between myself and military policy-makers by translating academic insights into frames that could be understood by commanders. The effect of my research and teaching activities on the IDF was thus filtered through the networks and alliances of military behavioral scientists.

Given this situation, the influence of my work can be captured by a metaphor of reverberation wherein my impact is not direct but rather is one that has secondary and tertiary effects. By this metaphor I mean that while my work did have some direct influence on my military allies, it was actually via their actions that the impact reverberated out through the military. Yet, at the same time, it is hard to deduce the influence of one action (or even one kind of action) on parts of the armed forces. Rather, it was the sum of my activities – including lectures, supervisions, publications, or committee work – interacting with the actions of others that sometimes led to change.

But there is one more point here regarding the "fuzzy" model. Many of the research-related activities I carried out – variations of research projects, brain-storming sessions, supervisions, committee work – served (again, with the actions of others) to create "special" times and spaces away from the often-hectic everyday military life. These spaces and times are crucial, I think, for finding the most fruitful critical distance from our understandings (as scholars and as troops) of the armed forces. In these encounters, what was often created were spaces to defamiliarize the assumptions, practices, and ideas that characterize the IDF or parts of it and thus allowed, at times, new formulations for understanding military behavior, and insights into the assumptions military people and academics work with. I call the kind knowledge produced in these interactions a "potentiating knowledge." Herein perhaps lies the real power of social research – it has a generative potential but whether change will occur is dependent on a host of other variables.

Conclusion: potentiating knowledge

As I have shown, the projects I participated in had a variety of formal aims such as undertaking research on combat companies, improving checkpoints, creating a new legal framework for reserve duty, or charting out how the IDF's units manage diversity. I now see that while there was relatively little direct impact of my activities on these matters, my influence lay much more in what Moskos (1988) calls an enlightenment model of military sociology: one of enriching the knowledge of soldiers and primarily officers (also Fosher, 2013). In terms of my contribution to critical knowledge (Levy, 2010, 2015) – one oriented toward starting public discussions about military-related issues and monitoring how the IDF manages organized violence or uses troops for other goals – my role was of a more participatory rather initiatory character.

Taking a step back, I can now ask again: what did my research-related activities in regard to the IDF "do"? As explained, an answer to this question involves going beyond each project's formal aims to the informal aims and sometimes hidden agendas of myself and other actors. As I tried to show in this chapter, my actions variously served to legitimate policy, ornament proceedings, further the careers of officers, open up avenues of research, add a voice to public discussions, lend academic credibility to in-house research, or set academic standards within the military's behavioral sciences units. Some of these effects were intentional while others were unintentional or only dimly realized at the time. Beyond all this, I now realize, perhaps my most important contribution lies in constantly attempting to create "potentiating knowledge", one that could be used by the IDF, the academe or parts of the general public.

References

Ben-Ari, E. (1989). Masks and Soldiering: The Israeli Army and the Palestinian Uprising. *Cultural Anthropology* 4 (4): 372–389.

Ben-Ari, E. (1991). *Changing Japanese Suburbia: A Study of Two Present-Day Localities*. London: Kegan Paul International.

Ben-Ari, E. (1998). *Mastering Soldiers: Conflict, Emotions and the Enemy in an Israeli Military Unit*. Oxford: Berghahn Books.

Ben-Ari, E. (2005). Journeying Out from the (Jewish-Israeli) Center: Research on Violence, Understanding and the Israeli Military. In J. Ramoneda, P. Villanova, W. Salem, & E. Kaufman (Eds.), *Breaking the Wall: The Social Responsibility of Palestinian and Israeli Academics and Intellectuals at Times of Violent Conflict: An Introspective Search* (pp. 161–168). Barcelona: Center of Contemporary Culture.

Ben-Ari, E. (2011). Anthropological Research, and State Violence: Some Observations of an Israeli Anthropologist. In L. A. McNamara, & R. A. Rubinstein (Eds.), *Dangerous Liaisons: Anthropologists and the National Security State* (pp. 167–184). Santa Fe: School of Advanced Research Press.

Ben-Ari, E. (forthcoming). Military Sociology in Israel. Forthcoming (2021) in M. Steinbrecher, H. Biehl and M. Elbe (eds) Empirical Research in and on the Armed Forces. Potsdam: Bundeswehr Centre of Military History and Social Science.

Ben-Ari, E., Lehrer, Z., Ben-Shalom, U., & Vainer, A. (2009). *Rethinking the Sociology of Combat: Israel's Combat Units in the Al-Aqsa Intifada*. Albany: State University of New York Press.

Ben-Ari, E., Maymon, M., Gazit, N., & Shatzberg, R. (2004). *From Checkpoints to Flow-points: Sites of Friction Between the Israel Defence Forces and Palestinians*. The Truman Research Institute for the Advancement of Peace. Gitelson Peace Paper.

Ben-Ishai, O. (2017). *From "Decision" to "Resolution": Change in the Doctrinal Language of Israel's Security Discourse as a New Legitimization for Military Action, since 2010*. Ben-Gurion University. Doctoral Dissertation.

Ben-Shalom, U., & Fox, S. (2009). Military Psychology in the Israeli Defense Forces: A Perspective of Continuity and Change. *Armed Forces and Society* 36 (1): 103–119.

Fosher, K. (2013). Reflections on Current Research: Science And Scientists In Military Organizations. *Journal of Culture, Language, and International Security* 1 (2) (accessed 24 June 2016).

Lehrer, Z., & Amram-Katz, S. (2011). The Sociology of Military Knowledge in the IDF: From 'Forging' to 'Deciphering'. *Israel Studies Review* 26 (2): 54–72.

Levy, Y. (2011). The Decline of the Reservist Army. *Military and Strategic Affairs* 3 (3): 63–74.

Levy, Y. (2015). Time for a Critical Military Sociology. *Res Militaris* 5 (2). http://resmilitaris.net/index.php?ID=1022274 (accessed 15 July 2020).

Moskos, C. (1988) *Soldiers and Sociology. SLA Marshall Lecture*. Washington, D.C.: US Army Research Institute for the Behavioral and Social Sciences.

Rein, Y., & Ben-Ari, E. (forthcoming). The Israel Reserve Law: The Duality of Reservists and Transformed Military Autonomy. *Armed Forces and Society*.

Sher, Y., & Ben-Eliyahu, H. (2011). Looking Out from the Tent: Internal Social Research at the IDF. *Israel Studies Review* 26 (2): 99–108.

Sørensen, B. R., & Ben-Ari, E. (2019). Rethinking Civil-Military Connections: From Relations to Entanglements. In B. R. Sørensen & E. Ben-Ari (Eds.) *Civil-Military Entanglements* (pp. 1–20). Oxford: Berghahn.

4 Bridging two worlds: the knowledge-policy linkage in the study of the Portuguese Armed Forces

Helena Carreiras

Introduction

Social scientific research on the military has included frequent reflexive quests regarding its cognitive dimension, focusing on the evolution of research topics, and paradigms and on *scientific impact* (Carreiras et al., 2016). Such efforts have been mainly directed toward mapping theoretical and methodological frameworks, as well as identifying core concepts, models, and tools used by the researchers (Lang, 1972; Harries-Jenkins & Moskos, 1981; Kümmel and Prüfert, 2000; Boëne, 2000; Callaghan & Kernic, 2003; Caforio, 2006, 2007; Kestnbaum, 2009; Ender & Gibson, 2005; Malesevic, 2010).

However, during the second decade of the twenty-first century, interest in the research process dynamics in military studies received a sudden boost. Different volumes that represented a 'reflexive turn' in the study of the military were published, raising questions from the point of view of the positionality of researchers, their engagement with the military and the social and methodological conditions for research in this field (Ben-Ari, 2011, 2016; Carreiras & Castro, 2013; Soeters, Shields & Ritjens, 2014; Williams et al., 2016; Carreiras et al., 2016). Few, however, have delved into the concrete question of the *social and policy impacts* of research, which is the object of the present volume. The knowledge-policy linkage, in particular, has remained largely unexamined.

And yet, while having its own autonomous sphere, scientific knowledge in military studies, as in other areas, has been considered fundamental to inform policy choices, by allowing the correct diagnosis of problems and the identification of resources and opportunities, promoting the critical evaluation of the conditions and effects of policies as well as the drawing of realistic scenarios.

But is this always the case? Which conditions foster or hinder that virtuous relationship? What kind of pathways to impact are there, and how do these relate to different modes of engagement between researchers and research stakeholders and end-users?

DOI: 10.4324/9781003165217-4

This chapter addresses these questions, by resorting to empirical information related to studying gender integration in the Portuguese armed forces, the recruitment and retention of contract soldiers, and the evaluation of military academies' programs by the National Accreditation Agency. It also offers a reflection on the role of the researcher in each case and the way it evolved through time.

Scientific knowledge, social impact, and the military

In contemporary societies, scientific knowledge has become a structuring component of most societal institutions, culture, everyday life, and social change. By assuming an ever-growing complexity, highly formalized and codified forms of knowledge have been incorporated into social processes, in ways well captured by the concept of Knowledge Society. Sociological research has been contributing for the understanding of the modalities in which scientific knowledge unfolds, the social conditions of its generation, transmission, and diffusion, its protagonists and mechanisms, its impacts, or the social dynamics it encompasses in processes of social change.

While the relation between scientific knowledge and its social uses or effects has long been a central topic for a variety of disciplines, from the philosophy of science and epistemology to the history and sociology of knowledge, the question of its policy impacts has acquired great prominence in recent decades. Not only has the concept of research impact pervaded academic discourses (Alla et al., 2018), but national assessment programs and funding agencies are placing increased emphasis on dissemination and impact evaluation. Impact has been defined in multiple ways, but most users would agree that it relates to some demonstrable and beneficial change in behaviors, beliefs, and practices, and there is an increasing demand for academic institutions and individual researchers to demonstrate the practice and policy implications of their work.

However, criticisms have been growing with regard to a frequently a-critical and dogmatic focus on research impact.

The first criticism is anchored on an epistemological call to resist the devaluation of the so-called 'fundamental research', emphasizing, on the contrary, the extent to which intellectual freedom and curiosity have been powerful drivers for progress. The pressure by funding agencies on researchers is considered to have a negative effect on the promotion of innovative and transformative ideas in favor of producing institutionalized knowledge. According to this perspective, many of the most interesting applied research results would never have been achieved without an environment that promoted creativity and set scientists relatively free from direct social and political agendas. Abraham Flexner's (1939) article on *The Usefulness of Useless Knowledge* has been often recalled to praising intellectual curiosity as the main driver for science (Flexner, 1939).[1]

The second criticism, which comes from impact researchers themselves, concerns the methodological limitations of current impact research. It focuses on the excessive emphasis on quantitative impact assessment and the lack of more qualitative studies. Some argue, for instance, that

> there has been too much emphasis in the current literature on the commercialization of research results, (…) quantitative indicators are not able to capture the whole array of social and cultural impacts of research. Therefore, more qualitative research on the impact of university research should be conducted so as to provide historical and discipline-specific accounts that acknowledge the long-time span behind important societal developments that intermingle with academic research. (Esko et al., 2012, p. 17).

Finally, a third criticism regarding most existing impact research is that there is a mistaken assumption of direct links – even if mediated through time and space – between knowledge and social action. Conversely, policy options have their own rationality, which is irreducible to the available knowledge in a certain historical moment. Policy decisions involve value and ethical choices that are multiple and often contradictory. According to this perspective, "no science choses for us" and there can be no deterministic or univocal connections between knowledge and policy.

Yet, it is important to avoid the trap of a situation, whereby upholding the values of independent, 'useless' knowledge, in Flexner's sense, would imply the dismissal of applied research or impact evaluation; or, on the contrary, one whereby focusing blindly on impact would restraint the space for free and independent research. Addressing social and policy impact is relevant in that one of the most important challenges societies face is the creation of new knowledge that can be useful for defining policies that help societies to advance (namely to be more sustainable, equal, and fair). In that sense, research must be accountable to people's and society's problems. Besides, it has also been noted that this new pressure is not only coming from top-down accountability policies. It also comes from below, from social demands that contribute to the discussions and arguments concerning the values and goals of science and policy. Even if there is no direct univocal link or single pathway between one and the other, their entanglement and the concrete ways in which this process takes place should be promoted and scrutinized.

Therefore, what is interesting here is to identify and analyze the modalities of the knowledge-policy linkage in concrete cases.

In this respect, the military provides an extremely challenging instance for scrutiny. Military institutions have been considered a privileged observatory for a variety of social dynamics, due to its specific features, which sometimes allows for enhanced visibility of the processes under study. This has been shown with regard to the way hierarchy and discipline stand as structuring principles that pervade military institutional dynamics in a rather unique way if compared to other bureaucracies (Castro, 2013) or the gendered nature of the military's prevailing inequality regime (Carreiras, 2017).[2] It can thus be viewed as a *strategic research site.* According to Merton, these are one type of *Strategic Research Materials* "that exhibit the phenomena to be explained or interpreted to such advantage and in such accessible form that they enable the fruitful investigation of previously stubborn problems and the discovery of new problems for further inquiry" (Merton, 1987, p. 1).

However, there are multiple pathways to impact and varying engagement modes between academic researchers and research end-users outside of academia. My own experience provides examples of these multiple pathways, of direct and indirect influences, and of the complex and often contradictory dynamics set in motion during civil-military encounters. I will thus revisit my work on gender integration in the armed forces, my involvement in an MoD project related to recruitment and retention of contract soldiers and my contribution to, and evaluation of, military high education programs. All these experiences illustrate the multiple trajectories of entanglements between civilian and military spheres (Sørensen & Ben-Ari, 2019), and between knowledge production and policy impacts.

Studying gender integration in the Portuguese Armed Forces

Indirect types of impact can relate to common situations when research aims at producing understanding and awareness, without directly aiming at policy outputs. My early scholarly work on gender integration in the Portuguese Armed Forces provides a good example of this type of mediated impact.

During my academic career, I conducted two large research projects specifically focused on gender integration in the Portuguese armed forces, both funded by the National Science Foundation and ignited from the academic context. The first one coincided with the start of gender integration in the Portuguese Armed Forces in the early 1990s. It was based on a mix-method design, combining an extensive survey to the first one thousand women to join the forces (Navy, Army, and Air-force) and in-depth interviews with a sample of male and female officers, NCO, and privates from the three branches. The second, conducted ten years later, was a larger comparative research on gender integration policies in western militaries, within the framework of my doctoral thesis. Both projects required a significant level of engagement and direct interactions with the military institution. I had to obtain formal authorizations to collect statistical data and documents, visit military locations throughout the country, or to interview military personnel. However, the projects' objectives did not directly aim at problem-solving or specific policy developments. They could be classified as pertaining to an *enlightenment model* of knowledge, with the main objective of furthering understanding of the social processes in motion. Still, both granted me access to the armed forces, as well as institutional trust and public recognition, on the basis of which I eventually developed a reputation as a respected scholar in the field, an authoritative voice among feminist audiences, and a trustworthy partner in military research-related activities. A variety of iterative activities emerged from this work that went far beyond its scientific impact, and I believe contributed to the specific understandings of the topic of gender integration among interested stakeholders, including future military leaders and commanders. I began to be regularly invited to give conferences and classes in military academies and war college promotion courses, supervise its students, participate in juris and evaluation committees, or publish in military journals.[3] My work has been extensively used and quoted in officer's thesis and articles on the topic, and I have received private and public recognition on different occasions, including invitations to keynote addresses at women's day ceremonies and similar events offered by both the Ministry of Defense and the different branches of the armed forces. In sum, a network of contacts that contributed to the circulation of knowledge and experience that would anchor future engagements developed.

All these activities involved 'productive interactions'[4] (Spaapen & van Drooge, 2011), in that two of the main pathways through which potentially productive interactions manifest themselves, have been activated: direct interpersonal interactions between the researcher and societal actors and indirect interactions mediated through publications and other written media (such as evaluations or expert interviews).

However, none of the policy proposals I had occasionally put forward throughout the years in my articles (this was the case of the creation of an office for gender equality at the MoD or of a gender-related complaints mechanism outside the chain of command) found its way into policy, at least not until recently. There was no window of opportunity, no adequate 'national mood' for them to creep into the political agenda at this stage. Nor did I take up the task of becoming more active as a 'political entrepreneur', besides occasional media participations, to ensure they did. Still, at a later stage, the moment when some of those proposals came to light, coincided with the existence of defense ministers with a social science background. In one case, the compulsory participation of women in the national defense day,[5] from which they were previously exempted – a measure that I had been supporting in public debates – was finally upheld in 2010 by a defense minister who is a respected scholar in the field of History and International Relations. In another case, both a Defense Gender Equality Plan and an MoD Office for Gender Equality were finally created in 2019/2020 by a minister who also comes from a scholarly (and diplomatic) background. Spillover effects from the international context of the *Women, Peace and Security* agenda and external pressures from international organizations where that agenda has been valued, have played a role in this case. However, the fact that top decision-makers and their direct staff (some of whom I had contact with from the academic field), had been exposed to, and were knowledgeable of, social scientific research, as well as of the public debates about the topic with larger audiences, had, in my view, a non-negligible impact on their sensitivity to propose and implement those measures, some of which did not enjoy the open approval of military elites.[6]

In sum, although none of my research proposals on gender-related processes had been developed in a policy-engaged mode, it is not implausible to consider that the ideas which it fed into the public debate found a sensitive environment in the academic-led ministerial offices. This variable, together with inevitable economic considerations, has certainly played its part in the final result.

This kind of indirect effects of research are particularly difficult to measure and, as the examples above show, there are many different pathways through which they might get to influence behaviors and policies. Besides, as Celso Castro's contribution to this volume shows, researchers cannot always control the effects of their work, which sometimes gains a life of its own. But even when that life is not easily trackable, a glance of the important effects it might have on research participants and other potential end-users of research is occasionally revealed to researchers. Something like that happened to me in September 2007 while I was participating in a conference on defense issues in Brazil. In that country, my Portuguese-written contributions on female military participation were already well-known. During a break between sessions, I was approached by one of the participants, a military woman from the Brazilian Air Force who was presenting the results of her master's thesis on female military participation. She thanked me for my work, saying that it had been very important for the development of her own research, and she added a sentence that still inspires and moves me: "your book helped me to realize that the problem was not me!". She referred to the way in which the study had highlighted the systemic nature of the problems experienced by military women and helped to dismiss naturalistic and individualistic explanations, which are so common, and, admissibly, so detrimental to the quality of women's military experience and organizational gender policies. The comparative dimension of the research had enabled her to understand the existence of regularities that condition attitudes and representations shared by military men and women in contexts of gender integration, which go beyond individuals and relate to cultural, organizational, and sociopolitical factors. This episode definitely reconciled me with the occasional doubt about the relevance or impact of my work and gave me the confidence that it probably had, and may continue to have, consequences. These were not only at the individual level, as in the example mentioned (which I would like to think empowered that woman to strengthen her agency), but also in terms of policies and social change, even if through secondary and tertiary impacts (see Ben-Ari, in this volume).

From *output* to *impact* in recruitment and retention policies

An important pathway to impact results from direct contract research in which specific agreements are established between researchers and the military and in which policy objectives usually ignite the research.

Between 2016 and 2019, following an invitation from the MoD's Personnel Directorate, I became the scientific coordinator of the project "Sociodemographic characteristics and organizational satisfaction of contract soldiers in the Portuguese Armed Forces". A formal partnership was established between my own research center at university, CIES-IUL, and the Ministry of National Defense (MoD), according to which I would be the scientific coordinator of the project, but with a clause that prevented me from using the data without the necessary institutional permission. The project developed in the framework of a larger MoD program aiming at monitoring recruitment and retention problems in the Portuguese Armed Forces, especially among contract soldiers.

Under my scientific coordination, a team that included a core group of sociologists from the MoD with whom I maintained long-lasting co-operative relations, as well as representatives of the Joint Chief of Staff, the Army, the Navy and the Air Force was gathered. The project drew on an extensive survey to all contract soldiers of the Portuguese Armed Forces, including officers, NCO, and privates from the three branches, aiming to study their motivations for joining, organizational satisfaction, and professional perspectives.

This was thus a typical contract research where the external scientific coordination was envisaged as enhancing the credibility and legitimacy of the project while simultaneously ensuring greater chances that its results would be made visible and given credit by both political and military top leadership. This was particularly important to the sociologists' team at the Personnel Directorate since there was a rather frustrating track record concerning a previous in-house research project. A similar survey had been conducted in 2008, which unraveled a variety of problems related to low motivation and eroding institutional affiliation on the part of contract soldiers, while at the same time pointing to degrading military service's material and training conditions. Unfortunately, that study had been kept in the drawer, without even being presented to the top leadership of the three branches in a closed discussion session, as envisaged by the research team as a minimal desirable output. Expectations were thus high among the MoD sociologists' team and me that this time would be different. In the eyes of the MoD team, my participation meant external, scientific legitimacy. But more than that, it was also expected to raise the value of the project's visibility and results, *internally*. To me, in turn, this was an exciting opportunity to co-produce knowledge that could feed policy decisions or help design realistic intervention scenarios. The team's belief was that the project would be vital for better understanding the context and determinants of the problem at hand, namely the difficulty in recruiting and retaining contract soldiers. Due to a variety of objective indicators of eroding service conditions that had been piling up, we all suspected that the situation in terms of motivations and institutional affiliation could have degraded even further if compared to the 2008 study. Additionally, we worked on the strong hypothesis that such erosion in commitment and motivation would relate more to those specific service conditions than to single variables related to salaries and allowances or even to external societal factors, as often voiced in official discourses. The possibility that research results related to soldiers' perceptions and evaluations could finally receive attention by decision-makers, and eventually have a concrete impact on policy, was a stimulating possibility.

The design and contents of the questionnaire, as well as the methodological issues of field research, were discussed and decided upon by the team during the first months of the project in a truly cooperative mood. At that point, this seemed to be a great example of civil-military research cooperation, holding promise of a 'shared ownership', which is crucial to raise the chances of effective impact, whenever different stakeholders cooperate. However, the different status of participants from the armed services – and their respective institutional location and linkage to top decision structures – did probably account for a varying degree of 'ownership' that at a later stage became obvious. In this case, the MoD was the contractor and the executive coordinator of the project. While the different armed forces military representatives were formally equal partners in the team, they were not researchers and their role was more that of being connectors or facilitators than of full-fledged team member. In any case, the fieldwork developed according to plans, with the necessary support from the three branches for the conduction of the questionnaire in the different military units throughout the country. Due to the complexity of the process and the difficult conciliation of timetables, it took longer than expected, but the first results could finally be shared in the summer of 2017.

The results revealed, as expected, a worrisome picture, even if it varied significantly between branches. The Army revealed the most difficult situation. Not only were soldiers very unsatisfied with an array of issues, ranging from material conditions to training or future labor market prospects, but also the data showed that institutional affiliation was eroding and that negative representations of the armed forces were growing the longer they remained in the ranks. A majority declared they would not recommend military service to friends or acquaintances. By the end of 2017, the project reports were ready and were internally disseminated, producing a shock wave throughout the military structure. The results were then discussed in a focus group interview with top generals and admirals, heads of personnel from the Navy, Army, Air Force, and the Joint Chief of Staff in February 2018. The goal of this focus group interview was that of gathering contextual information and strengthen the interpretative component of the study. But it had also been thought of as an additional instance to further the involvement of the military structure and hopefully raise the level of ownership and accountability among the various stakeholders. After this, recommendations for action were drawn and disseminated.

However, for over a year, both the reports and the recommendations were kept as 'reserved' by the minister's office at the MoD and the programmed public presentation was successively postponed. While the frustration grew among the MoD sociologists' team, I became aware that if nothing happened, my initial role as scientific 'legitimator' would gradually fade into an 'academic ornament' status (see Ben-Ari, in this volume). This was aggravated by the fact that I could not use the data to publish scholarly articles, and the few academic conferences in which those data were eventually discussed hardly served as compensation. It was only when a new minister and state secretary for defense took office following governmental elections in October 2018 that things started to move again. Following concrete requests for the results to be released (which I emphatically voiced), and a greater awareness of the study's relevance on the part of the new ministerial team, not only were the results publicly presented and discussed in April 2019, but they contributed decisively to a new 'Action Plan for Professionalization' issued in that occasion. Even if the political orientation of government remained the same, the characteristics of the incumbents (social scientists by training) seemed to have made a difference when it came to value social scientific results. Yet, was this enough to create impact? We finally had our main output disseminated (the project reports),[7] a valuable formal outcome in place (the Action Plan), but how, when, and under which conditions would the Plan be implemented, thus creating real impact?

The fear that this would become another plan 'in the drawer' grew, as two trends became clear: on the one hand, the lack of concrete monitoring of the plan's measures and timelines was already unveiling an implementation gap; on the other, there seemed to be a surprising detachment on the part of the different military services from the study and the Action Plan as collective assets. During a number of military internal and external public debate activities, representatives of the different services were not only referring to the study as 'the MoD's study' and totally disregarding their own involvement (and responsibility), by not referring to it at all, or preferring to present 'their own data', produced by the service-specific research units, when it existed. This means that at best, there was an *appropriation deficit* that resulted from a variety of possible causes,[8] but certainly not irrelevant for the sake of effectively creating impact.

Notwithstanding, the MoD designated a monitoring commission headed by a respected soldier-scholar, a retired army general, which again included representatives from the different services. There are of course many variables that can still interfere with implementation, but there is at least one more instance where appropriation can take place, which is in and of itself a positive development in a currently ongoing process.

Contributing to higher military education and evaluating it

One of the most common, if indirect, pathways to impact of research on the military has been the engagement of civilian researchers and scholars through teaching and supervision in military academies, or participation in oversight bodies of the military higher education system. Like many other colleagues, my experience includes a number of such participations. I have been frequently invited to lecture and give conferences in military universities, to participate in evaluation committees in officers' promotion programs, to contribute with forewords to publications of their thesis or workshop reports, or to meet with officers in exploratory interviews during the preparation of their research. But I have also followed the development of the Portuguese military higher education system from other observation and monitoring platforms. I was, for a short period, a member of the Council for Higher Military Education, an oversight body created in the Ministry of Defense, and of the Military University Institute Council, a consultative body of the post-graduate institute of the Portuguese armed forces (IUM). Finally, I have been part of an evaluation team of the Portuguese Agency for Evaluation and Accreditation of Higher Education, mandated to evaluate graduate programs from the three service academies.

The full integration of the military higher education system into the country's higher education system has been accomplished after years of adjustment to the Bologna process[9] regulations, which implied profound changes in the structure of teaching and research in Military Academies. This was a complex, requiring, and sometimes painful process, but a clear choice taken with a view to raising the overall legitimacy of military higher education in a world where university diplomas have acquired vital importance for social and professional recognition. The challenges have been multiple. My experience with the National Agency for Evaluation and Accreditation highlighted a few of these in a clear manner.

One of the major challenges was to conciliate the specificities of the military with the requirements of a system of higher education that is supposed to have strong linkages to research and the development of research centers. The small scale and variety of subject areas in Portuguese Service Academies' research centers had not allowed to build critical mass (both in terms of personnel and scientific production) for the centers to be accredited by the National Science Foundation, which represented a significant weakness in the framework of teaching programs' evaluation (even if cooperation with other research centers was already common). The typology of military careers, which did not include an educational path, also made it difficult to deliver on one important evaluation criteria, that of faculty stability. Additionally, problems existed, even if to a lesser degree, with regard to another fundamental criterion, the qualification of faculty, required to hold PhD degrees in certain proportions according to the educational level taught. This problem had been tackled through the recruitment of civilian teachers with the right qualifications and of officers holding degrees granted by civilian universities, whose number had been growing. However, both situations rose specific problematic issues: in the core military programs of infantry, artillery or cavalry, both civilian and military professors were trained and worked in scientific areas that might not be considered core to the study cycle – still another important evaluation criteria (which, in itself, raises the problem of the classification of military sciences within the Agency's portfolio of scientific areas). On the other hand, some of the officers who were PhD holders, posted as faculty in the Academies in order to fulfill the system requirements, felt they had been somehow harmed in their careers, having lost the opportunity to qualify for more prestigious and rewarding command positions.

All these and other tensions were particularly clear during the evaluation of a PhD program proposal by IUM which failed to be approved in the first instance. Rejection came on the grounds of lack of critical mass and outputs in research, as well as inadequate curricular structure. A revised program was later subject to evaluation and approved by a different evaluation team. In the meantime, the frustration with the initial result was frequently voiced in terms that underlined a central paradox: complaints regarding the lack of sensitivity of the Agency and its evaluators towards the specificity of the military coexisted with the valorization of a full and complete integration of military higher education into the national higher education system. This was expressed in frequent declarations of commitment to fulfill all common rules for relevance and accreditation (to be an 'equal' member of the system). Having been part of the first evaluation team, albeit not heading it, I was at times exposed to ambiguous statements on the 'incompetence' of evaluators, supposedly unknowledgeable of the uniqueness of the military system, but also on complimentary comments on the part of 'internal allies' who also questioned the initial proposal.

The autonomous development of this graduate level, within the otherwise fragmented and fragile research apparatus of Portuguese military institutions, might involve gains and losses. Will civil-military cooperation be strengthened, or will a more self-sufficient higher education military system promote isolation and entropy? This will probably depend on the type of engagements with other stakeholders within the national higher education and scientific systems that military academies and IUM will establish, fueled by both military internal needs and external pressure. Notwithstanding the pains in the process, the integration of military universities into the national higher education system, and its exposure to a common scientific culture, might create the conditions for greater impact of research within the military. If not for other motives, because it comes with the valorization of knowledge and reflexivity as fundamental tools for self-awareness and institutional change. This exposure and incorporation can be considered as yet another pathway to impact.

Conclusion

The experiences reported in this chapter allow us to highlight a few points regarding the linkage between research and policy and its expression in military-related contexts.

The first point is that there are, as in all other areas, multiple pathways to impact. In this case, examples of different modalities of engagement have been provided, focusing on activities and processes that involved *productive interactions* between the researcher (or the academic and scientific field) and the military through a) promoting understanding, b) co-producing policy, and c) creating context for potential impact. None of these engagement forms is free from tensions and paradoxes, which they might expose or even prompt to varying degrees. And different types of engagement may encompass different consequences or results. While formal engagements are required most of the time, informal interactions often appear as the most effective and productive.

The second is that there are great potentialities for civil-military cooperation when it comes to linking knowledge and policy, but this mutual implication is affected by *enabling or constraining conditions*. In the cases above described, enabling conditions were the preexistence of networking and trust, and supportive political leadership. *Constraining conditions*, related to the existence of non-homogeneous assorted coalitions of actors (see Ben-Ari in this volume) with different interests and intentions vis-à-vis the final result of the cooperation, claims for exceptionality, or tensions between the civilian (bureaucratic and political) structure and military actors.

The third point is the fact that there is a long and frequently delayed pathway from outputs to outcomes and then to impact. It is not always easy or possible to evaluate social and policy impact, even in light of identifiable outputs and outcomes of research. While outputs (publications, conferences), and to some extent, outcomes (new laws or regulations) can usually be assessed and even measured in quantitative terms, impacts (social and environmental improvements) are more difficult to assess, due to problems of causality and attribution, the complexity of the affected systems and the intrinsically qualitative nature of many impact categories (e.g., quality of military service, quality of gender integration).

Finally, and despite such difficulties, better impact assessment in military-related research will require the inclusion of clear indicators for monitoring and measuring impact – whenever that is the direct goal of research (as in policy-oriented projects) – as much as the encouragement of engagement modes between the relevant stakeholders that foster enabling conditions for more productive interactions in the civil-military realm.

Notes

1 Flexner's words keep echoing to underline the potentiality of knowledge beyond its direct 'utility': "The world has always been a sorry and confused sort of place – yet poets and artists and scientists have ignored the factors that would, if attended

to, paralyze them. From a practical point of view, intellectual and spiritual life is, on the surface, a useless form of activity, in which men indulge because they procure for themselves greater satisfactions than are otherwise obtainable. In this paper, I shall concern myself with the question of the extent to which the pursuit of these useless satisfactions proves unexpectedly the source from which undreamed-of utility is derived". (Flexner, 1939, p. 544).

2 Sometimes, however, and in a different sense, it might also prevent visibility. As noted by Woodward et al., "Conducting research on a closed institution structured around communications practices that assume secrecy and confidentiality has consequences for research (…), because it makes assessment of the effects of engagement and impact frequently difficult, and often impossible". (Woodward et al., 2020, p. 510).

3 Although publishing in military journals, many of which are not indexed in international databases is probably one of the most effective means to disseminate research results among participants and thus reach the professed goals of social and policy impact, the publication in non-indexed journals is irrelevant from the point of view of scientific impact.

4 Productive interactions are defined as the "exchanges between researchers and stakeholders in which knowledge is produced and valued that it is both scientifically robust and socially relevant" (Spaapen and van Drooge 2011, p. 212).

5 The National Defense Day is a one-day journey, including a visit to a defense and security-related unit, aimed at getting all 18-year-old young people acquainted with defense goals as well as the objectives, structure, and missions of the armed forces. The measure was aimed at equalizing the situation of both young men and women, with respect to the country's constitutional provision of non-discrimination related to gender.

6 The existence of a complaints' mechanism outside the chain of command is one of the less welcome proposals. On different occasions, in direct conversation with top-rank officers, they expressed the view that this kind of structure is not needed since the chain of command has the authority and ability to designate 'neutral' agents to tackle the problems of gender-related improper behavior and the like.

7 I could finally also write on the topic, as in the case of the chapter intitled "The Armed Forces are unable to attract and retain the professionals they need" published in 2019 in an important report *The State of the Nation and Public Policies 2019: Less Reforms, Better Policies* (IPPS-ISCTE).

8 Among these causes, there was, in my view, a corporative statement for autonomy and independence from the political structure on the part of the different services, through the valorization of their own resources, including some, even if limited, research capabilities.

9 The Bologna Process is a mechanism promoting intergovernmental cooperation between 48 European countries in the field of higher education. See https://ec.europa.eu/education/policies/higher-education/bologna-process-and-european-higher-education-area_en

References

Alla, K., Hall, W., Whiteford, H., Head, B., & Meurk, C. (2018). The Concept of Research Impact Pervades Contemporary Academic Discourse – But What Does It Actually Mean? *LSE Impact Blog*. https://blogs.lse.ac.uk/impactofsocialsciences/2018/01/09/the-concept-of-research-impact-pervades-contemporary-academic-discourse-but-what-does-it-actually-mean/ (accessed on 10 December 2020).

Ben-Ari, E. (2011). Anthropological Research, and State Violence: Some Observations of an Israeli Anthropologist. In L. A. McNamara, & and R. A. Rubinstein (Eds.), *Dangerous Liaisons. Anthropologists and the National Security State*. Santa Fe: School for Advanced Research.

Ben-Ari, E. (2016). What is Worthy of Study about the Military? The Sociology of Militaries-in-Use in Current-Day Conflicts. In H. Carreiras, C. Castro, & S. Frederic (Eds.), *Researching the Military*. London: Routledge.

Boëne, B. (2000). Social Science Research, War and the Military in the United States: An Outsider's View of the Field's Dominant Tradition. In G. Kümmel, & A. D. Prufert (Eds.), *Military Sociology: The Richness of a Discipline* (pp. 149–253). Baden-Baden: Nomos.

Caforio, G. (Ed.) (2006). *Handbook of the Sociology of the Military*. New York: Springer.

Caforio, G. (Ed.) (2007). *Social Sciences and the Military: An Interdisciplinary Overview*. London: Routledge.

Callaghan, J., & Kernic, F. (Eds.) (2003). *Armed Forces and International Security: Global Trends and Issues*. Munster: Lit Verlag.

Carreiras, H., & Castro, C. (Eds.) (2013). *Qualitative Methods in Military Studies: Research Experiences and Challenges*. London: Routledge.

Carreiras, H., Castro, C., & Frederic, S. (Eds.) (2016). *Researching the Military*. London: Routledge.

Carreiras, H. (2017). Gendered Organisational Dynamics in Military Contexts. In C. Duncanson, & R. Woodward (Orgs.), *Palgrave Handbook of Gender and the Military*. London: Palgrave.

Castro, C. (2013). Anthropological Methods and yhe Study of the Military: The Brazilian Experience. In H. Carreiras, & C. Castro. (Eds.), *Qualitative Methods in Military Studies: Research Experiences and Challenges*. London: Routledge.

Ender, M. G., Gibson, A. (2005). Invisible Institution: The Military, War, and Peace In 1990s Introductory Sociology Textbooks. *Journal of Political and Military Sociology* 33 (2): 249–266.

Esko, T., Tuunainen, J., & Miettinen R. (2012). Social Impact and Forms of Interaction Between University Research and Society in Humanities and Social Sciences. *International Journal of Contemporary Sociology* 49 (1): 17–46.

Flexner, A. (1939). The Usefulness of Useless Knowledge. *Harpers* 179: 544–552.

Harries-Jenkins, G., & Moskos, C. (1981). Armed Forces And Society. *Current Sociology* 29 (3): 1–76.

Kestnbaum, M. (2009). The Sociology of War and the Military. *Annual Review of Sociology* 35: 235–254.

Kümmel, G., & Prufert, A. D. (Eds.) (2000). *Military Sociology: The Richness of a Discipline*. Baden-Baden: Nomos.

Lang, K. (1972). *Military Institutions and the Sociology of War*. Beverly Hills: Sage.

Malesevic, S. (2010). *The Sociology of War and Violence*. Cambridge: Cambridge University Press.

Merton, R. K. (1987). Three Fragments from a Sociologist's Notebooks: Establishing the Phenomenon, Specified Ignorance, and Strategic Research Materials. *Annual Review of Sociology* 13 (1): 1–29.

Soeters, J., Shields, P. M., & Ritjens, S. (Eds.) (2014). *Routledge Handbook of Research Methods in Military Studies*. London: Routledge.

Sørensen, B. R., & Ben-Ari, E. (2019). Rethinking Civil-Military Connections: From Relations to Entanglements. In B. R. Sørensen, & E. Ben-Ari (Eds.), *Civil-Military Entanglements* (pp. 1–20). Oxford: Berghahn.

Spaapen, J., & van Drooge, L. (2011). Introducing "productive interactions" in social impact assessment. *Research Evaluation* 20 (3): 211–218.

Williams, A., Jenkings, N., Rech, M., & Woodward, R. (2016). The *Routledge Companion to Military Research Methods*. London: Routledge.

Woodward, R., Dawes, A., Edmunds, T., Higate, P., & Jenkings, K. N. (2020), The possibilities and limits of impact and engagement in research on military institutions. *Area* 52: 505–513.

5 In and between – tensions, dynamism, and practices

Soili Paananen

Introduction – impact in perspective

Impact is a concept that has many meanings. It is generally understood as how the research and its results are transformed into outputs, in other words, how successfully research has produced change. Detecting such a direct connection often requires metrics by which that connection can be assessed and verified. However, impact can also be understood as a broader concept that takes into account quality, and can create added value, or advance interaction between the military and society (Bartuševičienė & Šakalytė, 2013). It is precisely this broader meaning that I use in my efforts to describe the significance of social research, especially in the Finnish National Defense University context. Consequently, social research is reflected in the context of the Department of Leadership and Military Pedagogy with the focus on leadership, education, and related phenomena and the relationship between the military and society.

In my chapter, I explore the impact of social research in relations where the officer profession and academization, as well as politics, are intertwined. Professionalism in the military means that its practitioners have exclusive ownership of expert knowledge. Expert knowledge provides the ability to understand the operating environment and gives the power to determine problems and their possible solutions. However, at the macro level, professionals are provided guidance at the political level (Evetts, 2012). My main argument is that the priorities of these elements become visible in everyday practices as tensions. This creates a dynamism in daily reality that is sought to be balanced and resolved in different ways.

In the following chapter, I look at the impact of research from three perspectives and make use of the criteria set by the Finnish Education Evaluation Centre for the effectiveness of research when analyzing my own experiences. In the *Impact of research in research-based military education* and *Impact of research as societal impact* paragraphs, I look at everyday practices and situations in which tensions emerge and where there is an effort to balance dynamism. In the *Impact of research as innovation*, the focus is on research projects that have produced concrete change or have

DOI: 10.4324/9781003165217-5

the potential to do so. In innovations, the practices and means by which new knowledge could be put into practice become important. Together, these alternative ways of understanding effectiveness demonstrate different types and goals of knowledge. At the same time, they form a complex entity and are a matter for negotiations with various parties, both within the military and in relations between the military and the society (Ben-Ari, Carreiras, & Castro, this volume). However, before I move on to analyze the importance of research and scientific knowledge, I will briefly present the context of the work, which describes the Finnish National Defence University as part of the defense system.

The Finnish National Defence University

The Finnish National Defence University (FNDU) educates professional officers for the Finnish National Defence Forces and for the Border Guard. At the same time, it awards bachelor's, master's and doctor's degrees in military sciences and is a research-based university. FNDU offers students hands-on practical competences to prepare leaders for their duties and responsibilities during crises and states of emergency. It also teaches training skills, especially to young cadets, who after a bachelor's degree often serve as leaders in training conscripts in garrisons. This is based on the fact that the Finnish defense system is rooted in territorial defense, which is implemented through general conscription.

FNDU is part of the Finnish higher education system along with 13 other universities. As such, FNDU also undertakes the same academic tasks – research, education, and community relations – as other universities. FNDU has three departments: The Department of Warfare, The Department of Leadership and Military Pedagogy, and The Department of Military Technology. The main subjects comprise operational art and tactics, strategy, military history, leadership, military pedagogy, and military technology.

FNDU's education structure is similar to that of other universities, and therefore, it is administratively placed under the Ministry of Education. The bachelor's degree lasts for three years. After up to five years working as an instructor for conscripts, students can embark on a two-year master's degree course. After graduating, they usually go on to work as a unit commander in different units of the Defence Forces and the Border Guard. After completing their master's, officers have two options for further education: they can either take the Staff Officer's or the General Staff Officer's course, which leads to the highest-ranking positions. Doctoral studies are an option after the master's program for both officers and civilians.

FNDU operates also under the Ministry of Defence. Defense policy defines the guidelines and objectives of military defense. These policies are implemented by the defense command, which defines the requirements for the needed capabilities. This capacity building is attributed as a task for

FNDU and is resourced as part of the defense budget to be achieved. That way, FNDU implements defense as well as foreign and security policy principles that are imposed by political guidance.

Impact of research in research-based military education

Professional education vs. academic education

The impact of social research on the military can be interpreted as research-based military education (see "Armed Forces & Society," 2008) that is best described as academic education. Academic education means academization, in which academic standards have been adopted into military education (Libel, 2019). More specifically, this refers to teaching that is based on research, uses theoretical literature, and gives students the basics to both understand how scientific knowledge is constructed as well as how to carry out scientific research. However, alongside academic education, there are other issues promoted as part of military education and the military profession. Namely, the development of values such as honesty, responsibility, loyalty, and collegiality is held to be fundamental for entering the profession. The third important area of education focuses on developing mental and physical resilience. (cf. Jansen & Kramer, 2019, pp. 515–517.) For instance, in leadership studies, this means that although a broad theoretical approach is adopted on issues that are crucial for understanding military leadership as a whole, at the same time students learn practical leadership issues such as the importance of physical appearance to a leader's credibility. Finally, the officers' profession involves a lot of cultural knowledge, which brings out a wide range of significant issues that are part of their learning processes (Tiili, 2016). In sum, all these areas of expertise are generally advantageous for being military competent for their whole career. Hence, all these both academic and non-academic subjects are part of the everyday teaching and learning environment and cause tensions in relation to each other, especially when one thinks about how they are weighted.

In academic military education, theoretical knowledge also serves as the foundation for the practical aim. That is to say, military education and officers' careers are closely interconnected, and each stage of the education prepares them for the next level of their careers (cf. Jansen & Kramer, 2019). This is apparent, for example, in the leadership education program that is structured to progress gradually from concrete to more comprehensive leadership skills, and adapted to positions within the military that are equivalent to their achieved education level. After the BA degree the students work as instructors for conscripts. When studying for the BA degree, the students learn expert knowledge on leadership and leading in normal and emergency conditions. The objective is to learn to lead in both theory and practice. After taking the master's degree course, they usually go on to work as unit commanders in different units of the Defence Forces and the

Border Guard. Thus, during the MA degree, the nature of leadership changes to rationalized leadership, moment in which skills like delegating and following up become important. The time perspective also changes from the near future to the distant future, with the operational environment expanding from normal and emergency conditions to interagency cooperation between different authorities, crisis management, and peace support operations. Moreover, the learning objectives are extended significantly to provide a broad level of understanding on what leadership and leading constitute in different operational environments, and to enable students to conduct research into these matters. For those who complete their master's, there is a post-graduate General Staff Officer's degree. On the course interaction, communication, and cooperation in multifaceted and complex environments, in addition to organizational development, form the core of leadership education. At this stage, the ability to apply theoretical knowledge at a strategic level becomes a matter of importance (Paananen, in press). In short, theoretical knowledge is adapted in accordance with the education level and subsequent professional level. Academic education is expected to serve practical and professional aims as well as scientific understanding of militarily important issues.

To conclude, in military education, theoretical, professional, cultural knowledge, and practical skills are complementary, but they are highlighted differently depending on the level of education. More concretely, these are actual signs of how the Finnish military education is shaped by a long tradition. We see evidence of this, first of all, in the three-tier officer education system where officers acquire the knowledge and skills required for each leadership level, as I described in detail in the previous paragraph. Furthermore, other signs of that shaping by tradition are field manuals and military doctrines (Libel, 2019). An example thereof is the fact that field manuals for both leadership and education are used at the Department of Leadership and Military Pedagogy. They provide practical guidance, particularly on the leading and training of conscripts in garrisons. Deep leadership as a military doctrine also provides a leadership model that is taught at every level of education and whose principles are used to evaluate staff (see more later in this chapter). Thus, teaching, learning, and working practices that are built on historical and social processes (Billett, 2004; Devos, 2002; Engeström, 1987) still play a significant role in everyday learning and professional life. Hence, these practices also influence how the meaning of research-based knowledge is understood, appreciated, and implemented.

Abstract scientific knowledge vs. concrete knowledge for military needs

The impact of research can be understood as knowledge that improves the military and its functions. In fact, the stated purpose of the study is to produce concrete knowledge for military needs. This becomes apparent in the practices in which research topics are collected from across the organization. FNDU requests and gathers topics from the land, navy, and air forces, branch schools as well as from headquarters, university personnel, and the students themselves. With this practice, FNDU tries to address the challenges that have been identified in the field. Staff may also seek to integrate student theses into their own projects, allowing students to use their theoretical knowledge to guide theses while gaining valuable research information for their own projects. Students can also present topics themselves. They often relate to the issues they have encountered in their professional lives and which they want to examine in more detail in their theses, either motivated by pure intellectual interest or for purposes of their own future career paths. In brief, this practice seeks to utilize the academic research by focusing on topics that are relevant for the organization and to integrate that knowledge to develop it.

The same practice that is used for collecting topics for student theses also applies to the university's own research staff. The headquarters assigns research tasks to FNDU, usually long-term projects that last one to three years. These top-down research tasks will then be executed at the departments and the performance of the tasks will be evaluated annually. Quite often these research tasks are practically oriented. The aim of the studies is to measure, test, and evaluate specified issues in order to produce practical information for the organization and support future planning. Another defining characteristic of these tasks is that they are often focused on The Department of Warfare and on The Department of Military Technology. It follows that the topics to be studied are closely related to the development of material-technological military capabilities such as autonomous systems, sensor technologies, and defense products, in which social knowledge plays a minor role.

Besides top-down research tasks, researchers can make proposals for their own research projects. These projects will be submitted to headquarters for approval and subsequent assignments in accordance with existing processes. These bottom-up designed projects differ in many ways from the in-house created ones. First of all, these projects are usually carried out in cooperation with other universities, which, by definition, gives them a unique character. Another distinctive feature is that they are often financed by third-party funding; this, in turn, directs the focus of the research. A more important aspect is that the research topics are broad, multidisciplinary, and science oriented. Also, social knowledge can have its own significant role in these studies. This becomes apparent, for example, in a project funded by the Academy of Finland in which the purpose of the research is to understand the operating logic of the media, especially those related to terrorist events. In particular, the focus is on how terrorists disseminate their messages and how they use media for their own purposes. The most significant issue here is that the research can at best benefit both civil society and the military from a security point of view. Moreover, society can effectively utilize the central concepts of military science, such as strategic communication in its efforts to analyze electoral intervention, for example. (Huhtinen et al., 2019.) The challenge of such research is whether the military – or society – considers the knowledge significant enough to be linked to its processes, or whether it finds the areas where abstract and often indirectly connectable information can be implemented. Otherwise, if no concrete use can be found for academic knowledge, it is easily left out.

Researchers at the FNDU can also conduct independent research projects that do not go through the formal, administrative decision-making process. These projects are often linked to the researchers' own theoretical interests, have emerged from previous research needs, are related to the latest developments in science, or are based on a formed network on a common research topic. One example thereof is a project on complexity in crisis management missions. It began with in-depth interviews with military leaders sharing their personal experiences on being in crisis management operations. The interviews revealed extremely complex operating theaters in these missions. At the same time, complexity research was conducted at another university, but in a very different context. By combining theoretical knowledge on complexity with empirical knowledge, collaborative research could be created where the complexity of crisis management as a phenomenon could be made more comprehensible. When it comes to the impact of social research, the following can be highlighted here. Firstly, the complexity of crisis management is closely linked to vulnerable societies and to tensions between various religious, cultural, and economic communities. More specifically, it is about civil-military relations, which is one of the key areas of social research and has a considerable impact on analyzing the complexity phenomenon. More importantly, cooperating with scholars whose field of expertise brings in a new perspective increases efficiency and the impact of research. However, the challenge is that unless the research is in formal processes and is considered relevant, new knowledge may not be a subject, for instance, within the curriculum.

Science policy and its funding mechanisms also play an important role in defining the research topics to be promoted. In Finland, under the Ministry of Defence, The Scientific Advisory Board for Defence (MATINE) promotes research that supports military defense and security. In particular, it gives grounds for planning, decision-making, and development regarding defense capabilities. The board consists of nine committees, of which eight represent, in one way or another, "hard science" such as technology, electronics, materials, information technology, and medicine. Only one committee serves the social sciences. This broadly reflects the notion of a superiority of technical and material knowledge over social knowledge.

To summarize, the practice of collecting research topics that are important to the organization is significant when evaluating the impact of research. It is equally important to think about what kind of information is valued. In this case, technical and material knowledge, and in particular its tendency to be quantifiable, is essential in order to demonstrate its effectiveness. In comparison, social knowledge can appear vague and difficult to grasp. Furthermore, to ensure that the information does not remain unused, it is important to evaluate the practices and processes by which the researched knowledge can be implemented.

Impact of research as societal impact

Open vs. secret knowledge

Impact of research can also be understood as societal impact. Open research strengthens societal knowledge about military issues. Publications also provide a theoretical basis for both military and society to understand and act upon the phenomena therein. Thus, FNDU emphasizes the importance of open publications for its professors and researchers. Open publishing in journals, especially at a high – civilian and military – level, is strongly recommended nowadays.

In spite of its open publishing policy, FNDU is obliged to find compromises on research that serve also national and political interests. One aspect of this is that FNDU may be assigned research projects that are, by definition, secret. According to the Act on the Openness of Government Activities (1999), the secrecy is justified if the research touches on military intelligence, military equipment, military composition deployment or use, inventions, structures, systems, relevant items, or defense readiness if they jeopardize or harm the interests of national defense. Similarly, if it can be attested that theses, research plans, materials, technologies, and development work are detrimental, they may be classified as confidential (Act on the Openness of Government Activities, 1999/ 24.1§ 1–32). Thus, the law gives valid ground for research to be concealed if desired.

The tension between open and secret knowledge causes constant negotiations on whether theses and research reports are secret or non-secret. Firstly, theses, which are by nature public, are often the subject of negotiations on whether they are open, partially secret, or confidential. The most challenging cases are theses that use classified information. From the academic point of view, the information may be categorized as general knowledge according to scientific standards, but if the information comes from a classified source, it means that the theses must be either partially or totally classified. In addition, because the author is personally responsible for his or her written product and, in the worst-case scenario, liable for it before the law, the author usually ends up classifying the theses. Secondly, if the thesis is partially secret, it means that the written output has to be constructed differently compared to how research is usually written. More specifically, the author must separate the secret parts from the thesis and present them as separate attachments. Consequently, this means that special arrangements must be made for the submission of the thesis to the evaluator, so that it is not processed through the public network.

The same goes for theses openness. The public sections of the thesis are open information online, but the ability to read the work may be hampered by the fact that the theses may lack essential information concerning its comprehensibility. Finally, the parts of the thesis that are secret or partially secret cannot be checked by plagiarism prevention systems. At FNDU, the supervisors are required to scan the document with software to track duplicate content, but because the software compares text to online webpages and content databases, it automatically saves the content on the cloud. This is one reason why the requirement for open knowledge is not always fully met. As a result, the foregoing highlights the need for the organization to develop dual systems and practices to process knowledge according to the level of publicity.

External evaluations done by the Finnish Education Evaluation Centre (FINEEC) contribute directly to the fact that it is not possible to conduct only in-house research. FNDU is judged by FINEEC on the amount, quality, and impact of publications. FINEEC, which is supervised by the Ministry of Education, organizes periodic audits evaluating whether the quality of the work done at the university is in line with European principles. As part of this process, it is also assessed whether the knowledge produced by the FNDU is appropriate in relation to its strategy and whether this knowledge leads to FNDU's effective development. These external evaluations assess the impact of research and its knowledge production by harmonizing criteria and practices to be parallel with other civilian universities. Their public evaluation reports also outline the development of actions to be taken before the next audit. These interventions guide the direction in which FNDU should go and, at the same time, reflect and support the policy of civil society's understanding of effectiveness in research. These external evaluations indeed assess the credibility of FNDU's research on the basis of criteria used in other universities. This has a close link not only with the university's reputation to conduct management as a scientific actor, but also as a credible actor because it is funded through state budget. Even more importantly, FNDU must also enjoy the support and trust of society, because Finland's defense solution is based on general conscription.

In summary, military research has a social impact, but the organization must carefully balance its dual role. On the one hand, the national interest may demand the confidentiality of information; on the other, university status prescribes a public, scientific publication policy. The dual system creates constant tensions in everyday life and requires the organization to be creative in developing practices that allow these needs to be balanced.

Impact of research as innovation

Innovation vs. implementation

Impact of research as innovation is here understood as research results that can be used to bring about significant changes. The focus is on creating new ideas, using a new research methodology, or finding a new innovative perspective on a familiar issue. Innovation also produces something new with the future in mind. As a result, innovation may stimulate a process of translating research results into value for the military. In the following section, I will look at three different military cases that illustrate innovativeness in different ways, as well as the importance of research in the process of bringing about change.

Case 1: deep leadership model as a doctrine

The highly significant innovation implemented throughout the Finnish military organization is the deep leadership model. It was based on a need for change as the authoritarian way of leadership began to break down and, along with new ideas about leading people that began to emerge. The deep leadership model is based on in-house research, which was developed as part of Vesa Nissinen's (2001) dissertation, which subsequently became the official Finnish leadership doctrine. Deep leadership is a transformative leadership model in which leadership is understood as an interaction between people. It brings together two interconnected processes, namely leading people and deep learning about one's own leadership behavior. The model aims at enhancing leaders' interaction skills and their leadership skills by critically evaluating and developing their own behavior (Kazmi & Kinnunen, 2012).

The role of research in this fundamental transformation process has been crucial. In the beginning, Vesa Nissinen, as a teacher for leadership at the military academy (FNDU since 1995), began to develop the foundations for good leadership and its learning and development. The work was based on a gap in the curriculum, given that there were no courses on leading people or on leadership development. He started to develop ideas, tools, and metrics for a new leadership model with cadets. He also received data from Finnish war veterans that supported the emerging notion of leadership. At the same time, he completed educational studies at the university, from which the ideas of deep learning and lifelong learning emerged (V. Nissinen, personal communication, 3 June 2020).[1]

One of the reasons why the leadership model eventually achieved an official position in the military organization was, firstly, due to the fact that the model was based on research. Secondly, the creation of leadership profiles, their improvement with teaching, and the good assessments of teaching given by students established credibility for the notion that the model also worked in practice. Its reliability was also increased by the fact that the development work was in line with prestigious foreign military institutions. Finally, the legitimation of the deep leadership model also required individuals in top management positions to be convinced of the research-based nature of the model and, thus, resulting in its formalization as part of the entire Defence Forces' training system. However, the deep leadership model also received a lot of criticism and its implementation was opposed (V. Nissinen, personal communication, 3 June 2020).

Despite the criticism, the deep leadership model received positive feedback from the trainees and, due to its official position, it is now used throughout the organization, from conscript training to continuing education. Personnel, both civilians and officers, learn to reflect their communication and leadership skills and personal 360-degree profiles based on these skills carried out in the program. On the basis of this, people create learning trajectories to develop their communication and leadership behavior. So, in addition to developing a leadership model, the impact of research has been such that training programs, learning materials, and tools to evaluate and verify development as a leader have been developed to implement it. The personal changes produced by the training program can also spill over into people's work and operational environments (Kazmi & Kinnunen, 2012; Kinnunen, 2016; Kulmala et al., 2009).

Although the deep leadership model was created to correspond with military leadership training requirements, it has also been applied in civil society (Kazmi & Kinnunen, 2012; Kinnunen, 2016). In fact, a company, Deep Lead Inc, was established on the basis of the model to provide training and coaching for corporate and organizational leaders.

Case 2: the development of crisis management education

The PeaceTalk project, Talk and Interaction in Multinational Crisis Management Training, funded by the Finnish Academy,[2] studies talk and interaction in multinational crisis management exercises. It is conducted in cooperation with the National Defence University, the Finnish Defence Forces International Centre, and the Crisis Management Centre Finland. Education plays a significant role in preparing staff for challenging crisis management tasks. In crisis management, multidimensional and multinational situations require successful interaction and cooperation. Therefore, speech and interaction play a key role, especially when the working language is English. Speech and interaction are used to provide information, create situational awareness, resolve challenging situations, and negotiate between different communities and groups. With the help of the researched knowledge, crisis management education is developed (Haddington et al., 2019).

In this study, innovativeness, especially in the military context, refers both to data collection and to methods of analysis. Video recordings that simulate real-life practices and situations are collected in exercises, and the video data is analyzed in depth using conversation analysis as the method. This methodology provides authentic data from the exercises and can be used to assess how education helps participants to solve the challenges they face and to build a common understanding of the situations.

The impact of research can be assessed in terms of short-term and long-term effects. From a short-term perspective, research can produce visible phenomena in education, training, and thus actual operations, whose understanding is essential to the success in a crisis management operation. At the same time, this may highlight issues that require the re-evaluation of activities, the creation of new activities and practices, and more generally, an innovative approach in education in order to produce deep learning. The immediate impact of research findings can be seen in the teaching activities of educators as they become aware of the learners' challenges in simulated situations. With their advanced teaching skills, the research results can be transformed into better learning and action for those participating in training and exercises. From a long-term perspective, it is difficult to demonstrate the importance of individual training courses for the wider system, but their impact on the development of national defense can be assessed, for example. As Finland's security environment is increasingly interlinked with the international environment, the ability of the Defense Forces to utilize the expertise acquired in international crisis management operations and exercises, and the ability to apply it to the national environment, is of crucial importance.

Case 3: from the will to defend the state to defense relations

The will to defend the state is important in Finnish society because the defense system is based on a general conscription service. Every Finnish male citizen is liable for military service and must complete either armed or unarmed military service or non-military (civil) service. Women can apply for military service on a voluntary basis. The kind of relation that a citizen has with his or her own society also reflects his or her attitudes, perceptions, and actions toward military service and national defense. Therefore, analyzing this relation is significant because dealing with a crisis requires the support of all citizens.

The foundations of the Finns' willingness to defend the state date back to the 1950s, the period in which it began to be measured by an annual survey. That study has been based on a relatively limited understanding of the concept and has been quantified with a small number of questions. Its target population could also have been broader. Nowadays, this research perspective expanded and incorporated three studies that endeavor to bring out the perceptions of the whole nation about national defense and, thus, also about their own society.

The first research project analyzed citizens' relations to national defense. The study focused in particular on unconventional groups, such as non-military servicemen, conscientious objectors, and those who, having completed their military service, refused to stay in the military reserve (Kosonen et al., 2019). The research showed that citizens' relations with national defense is determined by their general attitudes to national defense, their trust in defense capability and defense politics, and know-how – that is, defense knowledge, abilities, and skills as well as agency, meaning one's own role in national defense (Kosonen, 2019). These findings showed a broader understanding of national defense, and based on them, a new theoretical concept, defense relation, was developed.

The second research project[3] extended the scope to citizens of all ages and in particular to women. The survey, which was theoretically based on the concept of defense relation, was broadly aimed at covering all citizens. In addition, the research involved interviewing women who participated in civic and NGO activities in the areas of security and national defense and, as such, observed training situations related to everyday security. The study aims to produce a comprehensive picture of the national defense and its relation to the entire population.

The third research project[4] analyzes what kind of information about national defense is offered and what kind of information the national defense relation is built on. The project specifically focuses on learning materials for upper secondary education, voluntary organizations, and programs aimed at the Finnish elite. In addition, the project also examines the information produced by the organization that has officially been responsible for measuring society's will to defend the state.

The innovativeness of these studies is twofold. Firstly, they developed a new concept, defense relation, which, compared to the previous concept of the will to defend the state allows for a broader spectrum of concepts of relations between citizens and society. Secondly, in addition to theoretical development, the research findings have the potential to be used in a wider societal debate. In fact, the results provide information on the development of the conscription service and the fulfillment of national defense duties, in particular in efforts to strengthen equality between all citizens. However, the issue is sensitive and strongly politicized. If desired, research can provide input for the reform of military or civic service. It also requires researchers to be actively involved in the societal debate, to use social media, and to be able to engage in decision-making bodies (T. Tallberg, personal communication, June 4, 2020).[5]

Discussion and conclusion

In this chapter, the impact of social research is understood as research-based education, as societal impact, and as innovation. Research-based knowledge is interpreted to develop education and assist an organization in its tasks, or as a two-way flow of theoretical knowledge between society and the military in an effort to investigate phenomena that are important to both, or knowledge that can be turned into change. All of these are associated with tensions that require practices and actions in an organization's efforts to balance with the dynamism created by them and to implement new knowledge. However, new knowledge that supports innovation and change may be resisted. Resistance is likely to arise from inside the profession.

Educators may also perceive the knowledge studied as a threat to their expertise and teaching competences, or the knowledge may be so politically sensitive that it does not enable preconditions for change. In situations where resistance exists, new knowledge challenges the existing one and becomes potential for change. It is important to note the importance of the new knowledge generated by the research also as a potential for innovation and change in both society and the military, because the demands for actionable and societally impactful knowledge are constantly increasing (Sannino & Engeström, 2017).

The military is a paradoxical environment where one learns to navigate a set of different demands and requirements (Jansen & Kramer, 2019). Similarly, the military seems paradoxical from a research perspective. It is accompanied by differing demands of politics, the profession, and the quality of scientific knowledge. An understanding of this contributes to moving away from the separation between society and the military and instead to seeing "the porousness between the two" (West & Matthewman, 2016, p. 489). Furthermore, by combining military sciences with other fields of science, rich and diverse information can be attained (cf. Jenkings et al., 2011) although there are challenges to be overcome. At best, information flows both within the military and in both directions between the military and society.

Notes

1 Nissinen, V. (3 June 2020). Interview by S. Paananen [Tape recording]. President of Deep Lead Inc. Helsinki, Finland.
2 https://www.oulu.fi/englishphilology/node/60783 (8 June 2020).
3 https://maanpuolustuskorkeakoulu.fi/en/tahto2-research-project (5 June 2020).
4 https://maanpuolustuskorkeakoulu.fi/en/research/maps1 (5 June 2020).
5 Tallberg, T. (4 June 2020). Interview by S. Paananen [Tape recording]. Professor of Military sociology. Finnish National Defence University.

References

Act on the Openness of Government Activities. (1999 / 24.1§ 1–32).

Armed Forces & Society. (2008), 35(1).

Bartuševičienė, I., & Šakalytė, E. (2013). Organizational Assessment: Effectiveness vs. Efficiency. *Social Transformations in Contemporary Society* 1: 45–53.

Billett, S. (2004). Learning Through Work: Workplace Participatory Practices. In H. Rainbird, A. Fuller, & A. Munro (Eds.), *Workplace Learning in Context* (pp. 109–125). London: Routledge.

Devos, A. (2002). Gender, Work and Workplace Learning. In F. Reeve, M. Cartwright, & R. Edwards (Eds.), *Supporting Lifelong Learning: Vol. 2. Organizing Learning* (pp. 51–63). London: RoutledgeFalmer.

Engeström, Y. (1987). *Learning by Expanding: An Activity-theoretical Approach to Developmental Research* (Doctoral dissertation). Helsinki: Orienta-consults.

Evetts, J. (2012). Explaining the Construction of Professionalism in the Military: History, Concepts and Theories. In D. R. Segal, & J. Burk (Eds.), *Military sociology* Vols. 1–4. (pp. 309–326). London: SAGE.

Haddington, P., Paananen, S., & Kamunen, A. (2019). Puheen ja vuorovaikutuksen tutkimuksen merkitys kriisinhallintakoulutuksessa ja kriisinhallinnassa [The Role of Speech and Interaction Research in Crisis Management Training and Crisis Management]. In *Puolustustutkimuksen vuosikirja* [Yearbook of Defence Research] (pp. 66–68). The Finnish Defence Forces.

Huhtinen, A. M., Kotilainen, N., Särmä, S., & Streng, M. (2019). Information Influence in Hybrid Environment: Reflexive Control as an Analytical Tool for Understanding Warfare in Social Media. *International Journal of Cyber Warfare and Terrorism (IJCWT)* 9 (3): 1–20. doi:10.4018/IJCWT.2019070101

Jansen, M. M., & Kramer, E. H. (2019). Military Mores – An Institutional Ethnography. *Ethnography and Education* 14 (4): 512–526. doi:10.1080/17457823.2018.1471612

Jenkings, K. N., Woodward, R., Williams, A. J., Rech, M. F., Murphy, A. L., & Bos, D. (2011). Military Occupations: Methodological Approaches and The Military – Academy Research Nexus. *Sociology Compass* 5 (1): 37–51. doi:10.1111/j.1751-9020.2010.00340.x

Kazmi, S. A. Z., & Kinnunen, T. (2012). Deep Leadership Coaching Effectiveness in a Corporate Scenario Constitutes Proactive Leadership Solution for 'Optimal Team Formation'. *European Journal of Social Sciences* 31 (2): 166–189.

Kinnunen, T. (2016). Deep Leadership Coaching Effectiveness. *Journal of Military Studies* 2 (1): 63–93. doi: 10.1515/jms-2016-0178

Kosonen, J. (2019). *Kansalaiset sotilaina ja maan puolustajina: asevelvollisten sitoutuminen maanpuolustukseen ja sen eri tehtäviin* [Citizens as Soldiers and Defenders of the Nation: The Commitment of Conscripts to National Defense and Its Various Tasks] (Doctoral dissertation). Publication Series 1, No. 33. Helsinki: National Defence University. Retrieved from http://urn.fi/URN:ISBN:978-951-25-3093-9/.

Kosonen, J., Puustinen, A., & Tallberg, T. (2019). Saying No To Military Service – Obligation, Killing and Inequality as Experienced Problems in Conscription-based Military in Finland. *Journal of Military Studies* 8: 46–57.

Kulmala, H. I., Ahoniemi, L., & Nissinen, V. (2009). Performance through Measuring Leaders' Profiles: An Empirical Study. *International Journal of Production Economics* 122: 385–394. doi: 10.1016/j.ijpe.2009.06.013

Libel, T. (2019). From the Sociology of the (Military) Profession to the Sociology of (Security) Expertise: The Case of European National Defence Universities. *Defence Studies* 19 (1): 62–84. doi:10.1080/14702436.2018.1562910

Nissinen, V. (2001). *Military Leadership: Critical Constructivist Approach to Conceptualizing, Modeling and Measuring Military Leadership in the Finnish Defence Forces* (Doctoral dissertation). Department of Management and Leadership. Publication Series 1, N:o 20. Helsinki: National Defence University. Retrieved from http://urn.fi/URN:NBN:fi-fe2016103125623/

Paananen, S. (forthcoming). Military Leadership Education and Training for Extreme Situations in Finland. In M. Holenweger (Ed.), *Training for Leaders in Extreme Situations*.Cham:Springer.

Sannino, A., & Engeström, Y. (2017). Co-generation of Societally Impactful Knowledge in Change Laboratories. *Management Learning* 48 (1): 80–96.

Tiili, M. L. (2016). Ammattilaisuuden ankkuripaikat: Kinesteettinen ja kulttuurinen tieto Suomenlahden merivartiostossa. [Anchoring professionalism: Kinesthetic and Cultural Knowledge in Finnish Coast Guard Units]. (Doctoral dissertation). Helsinki: University of Helsinki.

West, B., & Matthewman, S. (2016). Towards a Strong Program in the Sociology of War, the Military and Civil Society. *Journal of Sociology* 52 (3): 482–499. doi: 10.1177/1440783316654998

6 Research as a totem: How an academic book became the patrimony of a generation of officers

Celso Castro[1]

When Helena Carreiras and I proposed a session for the 2018 Congress of the International Sociological Association on the impact of sociological research on the military, we were trying to stimulate reflections on a subject that has not yet been extensively studied by the sociology of the military. Despite the vast literature and the increasingly reflexive mood in this field of study, reflections on how military institutions deal with the fact that they are the object of social scientists' research remain infrequent.[2]

In addition to this academic justification, I was also thinking, I must admit, on an opportunity for reflecting on a personal experience. Over three decades, but especially in recent years, I have accumulated several elements on the impact that my own research experience at the Military Academy of Agulhas Negras (AMAN), where the Brazilian Army career officers are trained, had on my "natives".

I refer to the fieldwork that I did at AMAN between 1987 and 1988 for my master's degree, the result of which I presented in 1989 in the Graduate Program in Social Anthropology of the National Museum (Federal University of Rio de Janeiro). It became a book in 1990, through a well-known Brazilian publishing house extremely important to the academic world: *The Military Spirit: a Study of Social Anthropology at the Military Academy of Agulhas Negras* (Rio de Janeiro, Zahar, 1990). The book dealt with the process of social construction of the military identity, along with their professional socialization.

When it was published in 1990, the press and the community of social scientists received the book with the brand of a novelty, of a pioneer study because it was the first of its kind made in Brazil by a social scientist. To this day, the vast majority of published works on the military in Brazil had not trespassed the walls of the institution to try to understand what Malinowski called "the native's point of view".[3] Generally produced in the field of Political Science, these studies resorted to secondary sources to deal with what was called, in the 1970s and 1980s, the military "ideology" or "mentality." The ultimate goal was, usually, to understand the behavior of the military in the political sphere.

It is important to note the historical context in which the research and

DOI: 10.4324/9781003165217-6

publication of the book took place, shortly after the end of a long military regime in Brazil, initiated with the *coup d'état* of 1964 and concluded with the transition to a civilian government in 1985. Because of this long authoritarian period, there was a high degree of mutual mistrust between the military and the academic world. Sometimes I felt to be in the worst of the two worlds. I was seen by some colleagues as a "suspect" for studying a theme that carried the mark of a "symbolic pollution". In the best version, I was warned to be careful, in order to avoid end up being tortured or killed in the fieldwork... However, at the same time, I was also seen with suspicion among the military. For them, at that time, sociology was often synonymous with socialism, and research, with espionage.

Looking back at these events three decades later, I can say that the reception and the impact of the book in the Brazilian academic world were very good. It was the first of a series of researches and publications that put me in the position – and I try to say this without false modesty – of a rare type of "specialist" in military sociology in Brazil. The book was extensively quoted, generally positively, by practically all social scientists (and also historians and researchers in related fields) who have since then devoted themselves to studying the military in Brazil.

In addition to its pioneer character, however, *The Military Spirit* remains one of the few studies of its kind on the Brazilian Armed Forces. I believe that the main theses of the book remain valid for understanding the training of Army officers even today, as they refer to long-standing social processes.

As for the reception and the impact among the military at the time, the answer was never clear to me. I sent a copy of the thesis to AMAN's command the day after my final examination. One year later, I sent invitations for the book's launch, held at a bookstore in Rio de Janeiro. Only one cadet appeared. He told me that several colleagues had planned to attend but could not do so because of a field maneuver scheduled for that week. He could go because he had been injured and was on medical leave. I gave him a book, and later sent two copies to AMAN's library.

Despite that, I never got an official institutional return on the book. I imagined that I was going to be invited by the commander of the Academy or the officers involved in the process of professional socialization of the cadets to discuss the book, but that unfortunately never occurred. Only in 2016 – 26 years after the publication! – a commander of the AMAN invited me to visit the institution and talk about it. Despite this long institutional silence, I have always been well greeted on the occasions when I returned to AMAN to do additional research or visits.

I believe that, to a great extent, this silence can be explained by the dynamic nature of the military career in Brazil. Usually the officer remains for a short period of time – not more than two years – in the same garrison. The geographic mobility along the vast national territory is a characteristic of the career, which aims at attributing a national character to the Army officialdom. When the book was published, many officers and most of the cadets with whom I had interacted during the research had already left AMAM – respectively, either because they had been moved elsewhere, or because they had finished the course.

This way, I was restricted to scattered opinions, taken at random, here and there. I learned, for example, that an Army Chief of Staff recommended his subordinates to read the book. Once, an officer told me that he was impressed when his girlfriend, who was very unaware of the routine of military life, suddenly began to comment on AMAN's daily life. Shortly after, he discovered that the "secret source" was *The Military Spirit*. At other times, officers praised the accuracy with which life at the military academy had been described. One day I met a cadet whom I had interviewed, and he said something that impressed me: "The book changed my life: I never called anyone else a *paisano*." *Paisano* is the derogatory term by which the military refer, among them, to civilians.

There were also negative reactions to the book, though some for curious and unexpected reasons. A lieutenant said to me he was bothered by the fact that I kept the slang and some bad words in the cadets' translated speech ("-What will civilians think?"). Other times I received indirect information that some military officers did not like the book, but that was sporadic and never "official" in the sense of being verbalized by some instructional authority.

Over the years, the book has become well known to the military, although it has not been read in the same proportion. I realized that for many officers, even without reading the book, the main concern was whether I spoke good or bad things about the Army, whether I was a "friend" or an "enemy". This can be a common concern for those who follow a career focused on a scenario of action (the combat) in which it is necessary to clearly distinguish friends from enemies. In part, too, the book's "career" was affected by a sense derived from the Brazilian historical and political context in which the institution was often criticized in the media and the academic world.

In recent years, however, I have come across several experiences that have led me to a new insight on the impact of the book on the military that I studied 30 years ago. I will briefly recount two of these experiences.

In November of 2015 I visited the Preparatory School of Cadets of the Army, located in Campinas (state of São Paulo) which functions as a headquarter for the first year of the formation course, to be continued in the following years in AMAN. We were warmly greeted by the colonel commander, who asked me when we met: "– Aren't you remembering me?" In response to my embarrassment, he identified himself as having been a cadet during my research at AMAN, and said that he had been interviewed by me. After my apologies for my decaying memory, he invited us to have lunch with him. During lunch, he talked animatedly and nostalgically about the AMAN of that time and about the book I wrote. However, when I asked about the year of his graduation, a problem arose. He was from the class that graduated in 1986, and he referred to that year as what he would have witnessed my research (and the supposed interview).

In fact, however, my fieldwork occurred in 1987–1988, not in the previous year. Therefore, we had not met at the Academy, let alone interviewed him. For a moment, I wondered whether I should undo the misunderstanding, but my researcher's curiosity spoke louder. I told him about the discrepancy between the dates, and that I had not been to AMAN while he was there. Gradually he realized that he had a "constructed memory." He was embarrassed and corrected himself by saying that from so much that he had over the years heard and talked of the research, and having read the book, he should have been confused to the point of thinking he had lived through the experience.

He was not, of course, lying. He had in fact indirectly constructed an individual memory of this supposed encounter, from culturally relevant milestones to his generation of officers. In that sense, the episode portrayed the solidification of a collective imaginary, which makes people who did not actually witness the research feel part of it.

A few months later, in April 2016, I was invited to give the inaugural lecture on the highest school in the military education system, the Course of Politics, Strategy and Senior Army Administration (CPEAEx). The class was a group of about 50 colonels who had potential to become generals. At the end of my presentation, I answered some questions they made related to the subject of the lecture. When the commander was about to close the event with protocolary thanks, a student asked for permission to make one last question, because he was sure that "everyone" on the auditorium wanted to do it: what did it meant for me to have participated in the FIT?

I was surprised by the question, which had nothing to do with the subject of the lecture. It referred to an episode of my fieldwork. In October of 1987, in the initial moments of my research, I integrated, wearing an uniform, a patrol of ten cadets of the first year of the AMAN course, in a war exercise called FIT, whose initials stand for Fiber, Initiative, and Tenacity. This exercise crowned the military instruction of the first year of the course, and as such, was eagerly awaited by the cadets. With an estimated duration of 53 hours, the FIT aimed to verify in practice what the cadets had learned during their first year: use of explosives, transposition of rivers and obstacles with ropes, use of compass and topographical map, camouflage, prisoner rescue, and instinctive shooting, among other war exercises. The trajectory between the ten "workshop points", where instructors waited for them to apply the practical exercises, had to be conducted in a combat situation, without patrols being seen, hiding from officers and from a supposed enemy.

I will not go into detail here about this experience, of which I participated actively almost to the end, in a climate of increasing complicity and companionship with my fellow patrol colleagues. I can summarize by saying that it was a turning point in my research. When, a few days later, I returned to the military academy for a new fieldwork season, I had a surprise. The cadets said, "there was only talk in the AMAN of the sociologist who had gone to FIT", and that the patrol had become "historical" because of the unusual participation of a sociologist. One said it was the "sensation of the moment", that "everyone" asked about my performance and that there were rumors that I would also participate in other military exercises. In an unperceived way at the time, participation in this exercise had been an important *rite de passage* for me. From there on the ceremonial manner with which it was treated by officers and cadets came to an end, and the fieldwork flowed swiftly.

Doing things together with the natives is not new in the history of field research in sociology and anthropology. By the late 1930s, William Foote-Whyte had rigged election votes along with the "natives" in the Boston neighborhood he was doing research to write *Street Corner Society* (1943). At the beginning of his famous "Deep Play: Notes on the Balinese Cockfight" (1973), Clifford Geertz tells how he and his wife, until then seemingly ignored and socially almost invisible researchers in a village in Bali, suddenly became the center of attention after running, along with the natives, to flee a police raid.

In my case, what surprised me at the time – and still surprises me when I recall these events and reread my field diary – is the unusual fact of my presence, as an apprentice anthropologist, researching in AMAN at that time and, in particular, running with the colleagues from "my" patrol. That fieldwork experience never happened again.

It is interesting to note that when a second edition of the book was published in 2003 – 13 years after its launch – the publisher suggested changing the subtitle from the savorless "a study of social anthropology at the Military Academy of Agulhas Negras" to "An Anthropologist in the Barracks", title of the final chapter of the book. The idea was to emphasize what had become the hallmark of the book: the unusual fact of a civilian doing research inside the barracks. On the cover of the book, an old photo of me in uniform, participating in the FIT.

Returning to the inaugural class at CPEAEx in 2016: why did that question come from the auditorium, 30 years after the fact? Surprised, I replied that I had lived the experience at that time more as a participant than as an observer, and only after that I realized how important it had been for me, as a *rite de passage*. The colonel who asked the question then revealed, excitedly, that two colleagues of "my" patrol were present in the auditorium. Therefore, a formal inaugural lecture ended in an atmosphere of relaxation and *communitas*, as Victor Turner would say.

For the last two years, I got in touch with some of my patrol cadets and talk to them about what we did together three decades. One of them remembered when he was warned that the patrol had been chosen to be accompanied by a civilian who was doing a research. It was a mixture of pride for the trust but also of concern, as he knew that FIT was not an easy exercise and that I could jeopardize collective performance. Another former participant of the patrol told me that, looking back, he found this experience and the book:

> Fantastic. Really sensational. It was as if you were a war correspondent. Recording our actions, decisions, mistakes and correctness. No make-up [...] It will always be a source of pride for my career to have had this opportunity.

Another one told me that those who participated in that FIT patrol were very much affected by the presence of the civilian, going through a hard time together with them. He said that this episode was remembered with some frequency, for example in the annual meetings of the class.

Several other less spectacular episodes of casual encounters with former cadets could be mentioned. However, I think these are enough to illustrate how, in an unanticipated way, my fieldwork and the book eventually became, many years later, a *patrimony* for a generation of officers, a kind of "totem", an identity symbol that serves as an emblem for the cadets or young officers of the time. Over time, the book became a patrimonial register of a time that had long been gone for them. This was neither formal nor official, promoted by the military institution, but a collective experience among a group of individuals. In a sense, the book ceased to be an inert object, becoming something embedded in a network of social relations.

At this point, it is interesting to remember a text by the anthropologist Ralph Linton on totemism and the American Expeditionary Force in World War I, written from his own experience as a member of the 42nd Division, called the "Rainbow". Linton intends to show the development, in the American Army in action in Europe, of a series of beliefs and practices that bear a considerable resemblance to the totemic complexes existing among some primitive peoples. He mentions how the name of the Division, arbitrarily chosen by members of the American General Staff, was gradually presiding over a growing sense of group solidarity. Initially, the soldiers began to answer the question "What unity do you belong to?" With "I am a Rainbow". Then, with the passage of time, the soldiers themselves established the use of a rainbow as a badge, respect for the drawn representation of this "patron", the belief in their protective role and their omen value – the appearance of a rainbow before a match came to be considered a sign of victory. This process was not confined to the 43rd Division: at the end of the war, the whole American expeditionary body was divided, according to Linton,

> into a series of well defined, and sometimes mutually jealous groups each of which had its individual complex of ideas and observances. [...] The individual complexes bond the members of each group together and enabled them to present a united front against other groups. (Linton, 1924:298).[4]

It is also important to highlight the *physical nature* of the book.[5] Something very different happens when the research results are presented orally, shared in pdf files, shown in slide shows or printed as articles in academic journals. Each of these document genres has different characteristics. The book is relatively durable (unlike ephemeral oral presentations), is accompanied by a front and back cover (unlike journal articles), can be lend or given as a gift as part of exchanges and so on. In this sense, the physical object – the book itself, and not just its text, transcribed elsewhere – possibly have a function analogous to the one Lévi-Strauss attributed to the *churinga*, objects of worship of the Australian aborigines that represent the reincarnation of an ancestor.[6] Like the churinga and our "historical" documents and objects, the book gives to a specific generation of Brazilian Army officers a physical existence to *their* history, embodies the intimate quality of the events, and puts it in contact with pure historicity.

Of course, *The Military Spirit* can also be seen as a mere "souvenir" of past times – an image less glamorous than the "totem" alternative. This image, however, seems to be more related to the individual's memory than to the collective memory of the group, and does not fully account for its dimension as a constituent element of its social identity.

In any case, we should abandon the conception of social identity as a "constant", something immutable that would "characterize" one group in favor of another conception, let's say "algebraic", inspirited by a passage by Georg Simmel, when he compares identity to a variable in an equation. Although it is always the same variable, its value changes according to the values of the other factors; in consequence, different results are produced.[7] This "algebraic" conception emphasizes the "political" dimension of social identity to the detriment of its "psychological" dimension. Identity ceases to be seen as a group property that is "assimilated" by its members and begins to be analyzed as the opposition between groups or categories that face both on the symbolic plane and on the level of social interaction.

In this sense, we might perhaps think of the object of this chapter as a case study of "civil-military relations." This expression has generated a quasi-autonomous field of study within the studies on the military institution, and for which I have no particular appreciation for two main reasons: it is based on a very American-centered point of view and naturalizes the existence of "civilians" or a "civilian world", when I prefer to view the "civilian" *as an invention of the military*.[8] In any case, the ethnography of how a simple academic book on the military has traveled in unexpected ways can make a contribution from the anthropological point of view to this field of study, extending its scope beyond that of the great normative conceptual schemes that dominate the field, and lead it to incorporate seemingly less grandiose but possibly more vivid nuances.

Let me conclude by making two general remarks. First, that we do not control the "career" of our work, that is, what they represent after they are made public. Second, that our works are condemned to historicity, even if we imagine doing research on the "present time". This is not to say that our work is "dated" – which they obviously are, in the sense that they are the fruit of a certain historical conjuncture – but that they *change* over time, taking on new forms but often unexpectedly.

Notes

1 PhD in Social Anthropology (Museu Nacional, UFRJ, 1995) and professor at the School of Social Sciences of the Fundação Getulio Vargas (FGV CPDOC), Brazil. celso.castro@fgv.br

2 An exception in the Brazilian case, although referring to the military police, is Piero Leirner's text, included in this book (see Chapter 7).

3 Again, an exception is the work of my colleague anthropologist Piero Leirner, Piero Leirner, who presented in 1995 his master's thesis, latter published as *Meia volta, volver. Um estudo antropológico sobre a hierarquia militar* (Ed. FGV, 1997). We organized together *Antropologia dos militares: reflexões sobre pesquisas de campo* (Ed. FGV, 2009). The book brought together nine Brazilian authors (and one Argentinean), who studied the Brazilian military using ethnography as a research method.

4 Linton, Ralph. 1924. "Totemism and the A:E.F.". *American Anthropologist*, N.S., 26: 296–300.

5 I am very grateful to Eyal Ben-Ari for pointing out this characteristic of the book, as well as the comment on the "civil-military relations" made ahead.

6 Claude Lévi-Strauss, *La pensée sauvage* (Paris, Plon, 1962), 315–323 (Chapter VIII, "Le temps retrouvé").

7 Simmel, Georg. 1977. *The Problems of the Philosophy of History* (p. 59), New York: The Free Press.

8 See my chapter in *Qualitative Methods in Military Studies,* ed. Castro and Carreiras (Abingdon, Routledge, 2013), p. 11: "The civilian is a military invention. I am only a civilian in relation to the military and when I am classified by them as such."

7 Notes on the appropriation of an ethnography: The case of the Military Police of the State of São Paulo, Brazil[1]

Piero Leirner

UFSCar, Federal University of São Carlos, Brazil

Although the issue of "appropriations of ethnographies" by groups has already been explored methodologically, it is remarkable that it has been rarely expanded to the reflection about the nature of the groups that make use of those means. It is more or less a "natural" datum that such ethnographic efforts are often aimed at groups considered "peripheral" (Bevilaqua 2000), a movement that culminated in a trend, already consolidated at this point (and widely discussed…), of "minority" or "non-hegemonic" groups producing their own "organic anthropologists" and their respective ethnographies.

What is also remarkable is that a large part of the debate involving the "limits of self-anthropology" has seldom been aimed at thinking about groups or agents as "central" or "hegemonic" as the anthropologists themselves, or more[2] (one exception is Latour [e.g., Latour & Woolgar, 1979]). And there has been even less conversation about cases in which such groups or actors employ or "appropriate" ethnographies. Of course, it can always be assumed that certain issues in the social sciences could become more or less "interested" – that a sociology of art could serve to legitimize emerging elites; that an "anthropology of anthropology" could make an effort to redefine the diagram of powers in the academic field; that there is what Pierre Bourdieu called "State science" (Bourdieu et al., 1994); that is, it is possible, if not expected, that an elite could even "buy" an academic work. However, along with those paths, there can also be the possibility of unexpected appropriations of ethnographies: those that get completely out of the anthropologist's control once their papers are published.

The present chapter addresses the description of the involuntary appropriation of an ethnography by a group that could be considered "central": officers of the Military Police of the State of São Paulo.[3] A caveat, first of all: it is true that such group differs quite a lot in *habitus* (Bourdieu, 1977) from those that could be called "elites" in our society.[4] But it is also true that Military Police officers are far from considering themselves to be sociological equivalents of factory workers, peasants, poor men from the periphery of the large Brazilian cities – just as that is certainly not how those segments see them.

DOI: 10.4324/9781003165217-7

It seems to me that the unexpected aspect of the situation resides in the fact that such "group" – and we are provisionally going to treat them as a group[5] – definitely does *not* need, at least at first glance, legitimizations such as those that anthropologists can provide, whether intentionally or not.[6] Well, maybe none of that is a matter of necessity; thus, let's put it another way: such group hardly has a taste for ethnographic descriptions, much less about themselves (at least, that is what I learned from their colleagues in the Army, during fieldwork).

The question that intrigues me is: why, then, was an ethnography of mine appropriated by such group, which used it as a model to think about principles that structure their own world? And more: why was I invited to the organization to see the results of such appropriation? The present chapter hence addresses the transfer mechanisms that took place after a series of events involving an ethnography.

A sequence of events

The facts that led me to learn about that appropriation refer back to the end of 2002. I was invited by a female officer of the São Paulo Military Police to go to the Barro Branco Academy, where the future officers of the organization are formed, for a ceremony partly based on data from my ethnography of the Brazilian Army, published in 1997, that had been incorporated into the curriculum of the Military Police cadets.[7] I later learned that incorporation resulted from the contact an officer instructor had with the book, after it was recommended by a fellow anthropologist that was teaching her a *lato sensu* post-graduate course at the São Paulo School of Sociology and Politics. I was at a loss to understand, because facts from years before led me to believe that my ethnography had not been well received in military circles. Thus, it is appropriate to clarify them.

Let us go back to 1995, when I was about to defend the master's dissertation that would originate that publication and found myself in the last steps of elaborating the ethnography, but was still in close contact with Army officers. In a history that is only worth outlining in a very synthetic manner,[8] I realized, after three years of what I had considered a series of frustrated attempts to carry out an investigation with members of the military, that throughout my contact with them – intermittent periods of contacts lasting between one day to two weeks – without realizing, I had been put face to face with a central principle of their cosmology (something similar to Nuer cattle, Azande sorcery, or Araweté maize). It was about hierarchy, a fact so obvious, so omnipresent in military life, yet that was obscured to the anthropologist, and also to the native that would like to understand what that anthropologist would actually like to study.

That same "invisible" hierarchy was the matter blocking what I was seeking then, the chance to synchronize military thinking with the Amazonian issue from "within". I was met with a certain silence about the issue, in contrast with a series of native teachings about what was asserted as the "true" point of view (global, cosmological, etc.). To be honest, at first, I paid little attention to that "point of view", which I understood to shift the center of the issue I would like to research. However, some incidents in the field gradually revealed that hierarchy was a fact that permeated not only the "internal" relationships and dimensions of military life, but also their "external" classifications, or their view of the "outside world".

The result of that is expressed in a thesis and a dissertation about military hierarchy, its implications for the construction of an internal world and its grammar within that which I later learned to be a "system of war" (Leirner, 2020a). That does not concern the matter at hand. What may be interesting is the fact that the organization responded to my statement that hierarchy is a "total social fact" for the military by "closing off" to future research. I never knew exactly why, despite speculating quite a lot. It is likely that although hierarchy is no secret (on the contrary, they use and abuse it), it is not admitted to be the *group's unifying principle*. As paradoxical as it may seem, what separates, unites... It is rather the basis upon which other principles manifest themselves, such as honor, discipline, patriotism, etc., which I considered to be the ideological glue of the hierarchical principle and which are taken as the true motivations that bind servicemen to the Army.

At the time, I also did not pay attention to a category the natives used insistently, but that years later it made sense to me. They used to say that a given country or person was a *friend* or an *enemy* of the Army. It is something generic enough to account for a person or a nation. I realized then that enmity was more than a mere word, but a modality of relation broad and deep enough to be handled with a higher level of abstraction than its everyday use might have suggested. Without digressing too much, I imagine this can be taken as a foundation to think about war in a more, let's say, "anthropological" sense[9]: war is a reciprocal and generalized relation of enmity.

Exploring the reasons and consequences of that conceptual use of war is beyond our scope here. For now, it is worth mentioning that someone (from the Army) told me that I was no longer a *friend* of the Brazilian Army; that is, at the very least (considering the best-case scenario), meaning I did not adjust to the plan that was initially expected for our relationship. I was supposed to be a sort of a *link* between the Army and the university, at a time (the 1990s) when those were extremely insubstantial. That did not happen: I became an anthropologist that called them "natives", and maybe that did not please someone...

About seven years later, I found myself invited, as a *friend* of the Military Police of São Paulo, to attend a ceremony that would pay a small tribute based precisely on the same research that caused me to be ostracized as a researcher of that "object" (keep in mind that it has much more power to decide who will and who will not research it). So, I attended the ceremony.

I went through a very well-known procedure. Someone was waiting for me, right at the entrance of the academy, which was supported by two large marble columns, in either a Greek or Ionic style, I do not recall which one exactly. That person took me to the person in charge of the whole operation – the instructor that had had contact with my book – and then she took me to the commander's office. That is standard procedure in military institutions: a subordinate receives you, transfers the guest to the interested or responsible party, who makes introductions to a superior officer – depending on the importance of the guest, it is either a high superior or a junior officer. In this case, I first went to the office of one of the people in charge of the course, and then we all went to the commander's office (it is worth noting that in that circuit, you always go through hallways, where you are introduced to most people going by, and from that point on, you go through anterooms, where an assistant announces you to the head of the department).

As I was conducted to the office, I noticed a meeting in which some Armed Forces commanders were present: of the Riot Police, the Federal Highway Police, the Fire Department, the Police, officers from the Barro Branco Academy, besides a judge, and if I'm not mistaken, a prosecutor. With the usual cordiality servicemen display toward their *friends*, I was welcomed, praised, questioned about what I was doing at the time, and was obviously made to listen to a conversation about the importance of initiatives aiming to acknowledge the "true" side of military organizations (which was also a prerogative of the Army). After that quick "living room", we all headed toward what really mattered: a large room turned into an auditorium, where the awaited event would take place.

It was a role-playing activity, a sort of military life "psychodrama" based on my book. It was the reenactment of a staging carried out as the graduation project of a group of cadets (that is, students of the course) which had been praised and received in a very positive way by the command of the academy. I did imagine it would be something of more intense proportions than I had anticipated, since at that point, I already knew what the presence of Armed Forces or Troop commanders meant. What happened then?

At the entrances, badges that only showed the ranks – soldiers, sergeants, colonels, etc. – were randomly distributed to the holders. That caused some embarrassment, from what I could observe: a colonel received a "soldier" badge and laughed nervously; I also got that one. Everyone took their seats in the corners of a room of approximately $80m^2$, following the hierarchical disposition (of the badges), as it is customary in military organizations: the highest-ranking officer sits at the center facing the door, and the subordinates distribute themselves in a descending order, one by one, to the right and the left of their immediate superior, respectively. Among the chairs, in strategic positions to fill symmetrical spaces of the room, people stood wearing hooded cloaks and holding candles, giving a certain air of mystery and a religious aura to the scene.

Once everyone was seated and in silence, a door was opened, where a young man apparently enacted, to the sound of music (I do not recall what it was, I think it was Beethoven), something that represented the series of stages a cadet goes through during his years at the academy: the frightened arrival; the hazing; the class camaraderie; the study routine; the tough physical training; the socialization. That stage lasted about five minutes. It was the preparation to the climax: when the student is about to give up and leave the academy, two marble columns reproducing precisely those that support the entrance of the academy are chained to his body: two miniatures named… Hierarchy and Discipline!

Then the young cadet enters the room, struggling to drag the columns, simulating crying and strong signs of emotions, yelling phrases such as "I can't do it anymore!", "I can't go on!", "I must give up!". Once he reaches the center of the room, at the same time he is about to get rid of the chains, the hooded people standing around the room remove their cloaks, revealing that under them was a representative of each command: Riot Police, Federal Highway Police, Fire Department, Police, etc. They go to the young man, hold him, lift him, raise the columns, free him from the chains, and make him realize that, deep down, those columns underpin his life in the organization ("and these columns will be my pillars from now on"…). The performance ends.

When the lights went on, I noticed a scene of commotion in the room; that certainly scared me, because I once again found myself in a situation in which I thought that the whole story about hierarchy and discipline was a great truism. But it wasn't, I was mistaken once again, and what is worse, I was mistaken based on my own work! They all scattered around, and the inverse circuit was made: I headed to the commander's office and was thanked for my services; I went down to the academy's facilities, where I was told that "*everything in this place is absolutely the same as in the Army*". I pondered: if everything is indeed the same, why was my ethnography received so differently? How can I be a *friend* here and an *enemy* there?

A conclusion, if that is possible…

Indeed, answering that question is only fully possible in a somewhat speculative tone. I would rather formulate a clue, based on a conversation I had as I was leaving the academy. Going through the entrance hall, I noticed that next to the columns there was a flyer with some things written about the course, and more specifically, about the class whose activity I had just watched. One thing caught my attention, a transcription of an excerpt from my book, without a reference, but which I knew was a passage selected from C. Lévi-Strauss' renowned "Introduction to the Work of Marcel Mauss" (2001 [1950]):

> To call the social fact *total* is not merely to signify that *everything observed is part of the observation*, but also, and above all, that in a

> science in which the observer is of the same nature, as his object of study, *the observer himself is a part of his observation*. (p. 29)

Why was that sentence randomly in the middle of other sentences such as "Knowledge, Ethics, and Conduct"? I remembered I always saw things like that in the Army, sentences nailed to walls, inscriptions randomly scattered around the landscape. A former officer told me it was a form of mnemonic thinking, principles that are internalized through minimalist and self-repeating references. I suspected that sentence had something to do with some "rule of twisting" of the original sense, and that the observer-observed relationship was being modulated as "friend-enemy". I couldn't resist the question, "If you are the same as the Army, so who are your *enemies?*".[10] There was no answer, simply because there could not be one. I knew that the logic of generalized enmity, for an army, leads to the limit-idea that indeed *all* other armies are potential enemies, that was why I was always told, "Piero, we are at war, because we are always deterring the *enemy*".

It is true the police can always claim to be deterring crime, but the fact is that armies are aware of who their enemies are and what they are like – and it is no accident they are all the same in their internal grammar (Leirner, 2020a). As to the police, I remain at a loss. Crime – and criminals – are definitely "different" and somewhat ethereal. Everything leads one to believe there is some confusion between what is a friend and what is an enemy for a Military Police, and then the idea that the observed is part of the observation (something that confounds the linearity of any information system!) began to make more sense. To go back to the beginning, we must see what the notions of *friend/enemy*, *observer/observed* can tell us about the nature of that group and its "appropriation" of the ethnography.

I have two things to say to conclude this paper: 1) it is not about a minority whose ethnography may "serve" as a legitimizing tool, nor is it about an elite that can also collect that merchandise as strategy for distinction in its extremely competitive world, since there is no competition here: it is a monopoly. 2) for that very same reason, nor was it about an "exotic" view toward the group itself: there was something we could call an appropriation as "functional reevaluation of the categories" (Sahlins, 1985): the content of the ethnography was absorbed and domesticated, transformed within the organization.

So why was it, differently from what happened in the Army? That leaves me with only one guess: the Brazilian Army, as well as any other army, is an organization aimed at the *outside*: the anthropologist's interpretation has everything to be seen as a horizon, if that is the case, to be absorbed "from the outside": it is an alien opinion. In the case of the Military Police, its position aimed at the *inside* of the society that it engenders; this allows external perceptions to be formulated *from within*... Albeit being two mirror-institutions, the only conclusion I can draw, in anthropology jargon, is that they are affine. Hence, my guess, based on Louis Dumont's old notion, is the following: an affine of an affine is consanguine, and so I stand...[11]

Notes

1 In Brazil, the Military Polices, according to the Federal Constitution of 1988, are not subordinate to the Armed Forces (except as auxiliary and reserve forces, in extraordinary cases such as the state of war). They are subordinated to the governments of the 27 states (and Federal District) of the Federation. Their function is ostensible policing and the preservation of public order. However, especially during the long Brazilian military regime (1964–1985), they were commanded by Brazilian Army officers and also assisted in political repression.
2 And "central" to the anthropologists themselves, we might add.
3 As will be seen ahead, I had contact with a group of officers, not with *all* the officers of the Military Police of São Paulo. However, given the event's official character and the hierarchical characteristics of the organization (see infra discussion), it is possible to somehow take the part for the whole.
4 Maybe they are rather a "group" at the service of the "center", but no less imbricated in it.
5 It is an organization with intense socialization mechanisms that aims, above all, to inculcate in its members the idea that they are a group apart from society, thus following orders very similar to those Castro (1990) observed among the cadets and I observed (Leirner, 1997) among Army officers. We will get back to that.
6 See Castro [1990], who introduced the anthropological analysis of the military in Brazil. For a comprehensive view of the various modes of interaction between academics and the military, see the introduction to this volume.
7 The cadets of the São Paulo Military Police enter the academy through a selection process carried out by FUVEST – the same institution that selects the students of the University of São Paulo. Until 2005, it was one of the three most sought-after careers offered by the examination, in terms of applicants to places ratio.
8 More detailed narratives about that "field research" can be found in Leirner (1997; 2020b).
9 Different, therefore, from the notion usually cited in the social sciences, making use of Clausewitz's axiom that "war is the continuation of politics by other means". A greater problematization of the concept of war in Anthropology, and of how it approaches a notion of politics, can be found in Leirner (1997).
10 If the reader is curious, it is possible to visit the Barro Branco Academy's website and see how it's similar to a typical military academy. Among its activities, there is even "territorial defense". http://www.polmil.sp.gov.br/unidades/apmbb/
11 A previous version of this chapter was published in *Cadernos de Campo* (São Paulo, n. 14/15, pp. 1–382, 2006). I wish to thank the editors of this book, specially Celso Castro, for the incentive to publish this text.

Bibliography

Bevilaqua, C., & Leirner, P.C. (2000). Notas sobre a Análise Antropológica de Setores do Estado Brasileiro [Notes on the Anthropological Analysis of Sectors of the Brazilian State]. *Revista de Antropologia*, 43 (2). São Paulo: USP.

Bourdieu, P. (1977). *Outline of a Theory of Practice*. Cambridge: Cambridge University Press.

Bourdieu, P., Wacquant, Loic J. D., & Farage, S. (1994). Rethinking the State: Genesis and Structure of the Bureaucratic Field. *Sociological Theory* 12 (1): 1–18. doi:10.2307/202032.

Castro, C. (1990). *O Espírito Militar: Um Estudo de Antropologia Social na Academia Militar das Agulhas Negras [The Military Spirit: A Study of Social Anthropology at the Military Academy of Agulhas Negras]*. Rio de Janeiro: Jorge Zahar Editor.

Latour, B., & Woolgar, S. (1979). *Laboratory Life: The Construction of Scientific Facts*. Beverly Hills: Sage.

Leirner, P. (1997). *Meia-Volta, Volver: Um Estudo Antropológico Sobre a Hierarquia Militar* ['Meia-Volta, Volver': An Anthropological Study on the Military Hierarchy]. Rio de Janeiro: FGV/Fapesp.

Leirner, P. (2020a).*O Brasil no Espectro de uma Guerra Híbrida: militares, operações psicológicas e política em uma perspectiva etnográfica* [Brazil in the Spectrum of a Hybrid War: Military, Psychological Operations and Politcs in an Ethnographic Perspective]. São Paulo: Alameda.

Leirner, P. (2020b). Hybrid Warfare in Brazil: The Highest Stage of the Military Insurgency. *Hau: Journal of Ethnographic Theory* 10 (1). Chicago: University of Chicago Press. doi:10.1086/708680

Lévi-Strauss, C. (2001). "Introduction to the Work of Marcel Mauss". In *Sociology and Anthropology*. London: Routledge.

Sahlins, M.D. (1985). *Islands of History*. Chicago: Chicago University Press.

8 The impact of social research on gender policies in the Argentine Armed Forces

Laura Masson
National University of San Martín

Introduction

The gender policies carried out in the Armed Forces in Argentina constitute a beacon light in South America. Currently, the percentage of women in the Armed Forces, excluding civilian personnel, accounts for 34.45% in the professional corps and 6.56% in the combat corps. The Ministry of Defense has a Gender Policy Directorate which reports to the National Directorate of Human Rights and International Humanitarian Law. The Army, Navy and Air Force each have a Gender Department, and there are 19 gender offices spread across the national territory and 11 teams for the attention and prevention of domestic violence. At present, all branches of the Argentine military are open to women; and female combat officers, who were the last to join, are advancing in their careers. These policies, aimed at promoting a cultural change in the Armed Forces, were based upon three spheres in which critical knowledge is built: social science gender research (particularly anthropology, sociology, and law); the tools provided by international human rights law; and the concepts and certain core matters of the feminist movement, mainly academic feminism and feminist NGOs.

In this chapter, I will conduct a reflexive analysis, inspired by positional analysis provided by Luc Boltanski (1973). I intend to explore my professional performance both in the academic field and in the State bureaucracy (as a gender researcher at civilian universities, lecturer in the Air Force Faculty, advisor at the Ministry of Defense, and Gender Policy Director at the Ministry of Defense) through the multiple institutional positions I have held. These multiple positions constitute a privileged point of view to understand the influence of social research on the military in Argentina, particularly because, as the proposal of this book highlights, the effects of such research imply complex processes of production, dissemination, and consumption of knowledge via dynamics of acceptance, understanding, interpretation, negotiation, assimilation, and implementation. In this regard, I take up the challenge proposed by the book editors to turn the question of the impact of social scientific research on the Armed Forces into an object

DOI: 10.4324/9781003165217-8

of study. Along these lines, I will discuss how social research impacts gender policies in the armed forces by analyzing four examples pertaining to my own career. Finally, I will make a brief reference to what happens when the researcher becomes a policy-maker himself.

This piece of research falls within the critical sociological model defined by Levi (quoted by Ben Ari et al.):

> this kind of knowledge-constituting interest is directed towards initiating debates to raise awareness of military-related issues (such as the management of diversity, gender-related issues, the effects of personnel policies, and civilian control) so as to monitor the ways by which States manage organized violence and use the military for other goals. Critical thinking identifies constraining structures of power, such as relations of dependence that unreflectively appear as natural

specifically "on how the Armed Forces entrench ethnic, race or—most important—gender relations" (Ben Ari et al., 2020).

Social sciences and the armed forces: a troubled relationship

According to Ben Ari et al. (2020) the military resembles any other public organization and yet possesses a uniqueness of its own. It is similar to large public bureaucracies in the sense that organizational change involves nonlinear processes in which multiple actors with multiple agendas (including those of researchers) participate, making it hard to measure or even appraise the impact of research. And yet, it is unlike these other large public institutions because its uniqueness lies in handling legitimate State violence. But, in some countries like Argentina (and others Latin American countries), where the legitimate State violence became illegitimate during the several military dictatorships and violence was exerted against the national population itself, this characteristic acquires a different meaning. In the Argentinian case, the world of the military and that of social sciences remained antagonistic for years. Therefore, exploring the impact of social research on the military requires knowledge of the relationship between the social sciences and the military in the contexts of Argentine history and context.

In the twentieth century, the country experienced six military dictatorships, namely in 1930, 1943, 1955, 1962, 1966, and 1976. The last one imposed State terrorism, which amounted to massive violations of human rights, hundreds of exiles, and about 30,000 forced disappearances of people. Lecturers and students of social sciences were counted among the exiles and missing persons. Social science study courses were severely affected and some of the faculties were even closed down. Therefore, for a long time the relationship between social sciences research and the military was tense and distrustful. In recent decades, most of the social science production has focused on the authoritarian character of the Armed Forces, the condemnation of the repression of their own people, the analysis of the modus operandi of the military dictatorship, and the construction of a collective memory about that "dark period" in Argentinian history. As a consequence, social science researchers who study another aspect of the Armed Forces, different from those mentioned above, were faced by colleague social scientists with suspicions of complicity with the military or with the danger of being co-opted by them.

The interest in investigating the Armed Forces as a public bureaucracy arose only in the first decade of the twenty-first century, almost 20 years after the restoration of democracy. One of the first authors in the field of military ethnography, Máximo Badaró, reflected on the dilemmas raised by the "anthropology of controversial institutions" (Badaró, 2009). One of his main dilemmas was how to establish and maintain proximity with his military interlocutors while keeping a fair distance at the same time. He reflects that the very limited anthropological work on the Latin American Armed Forces is related, to a certain point, to the different marks they left on institutions, ideas, and subjectivities. These marks were constantly present during my fieldwork and in the research process in general, and they manifested themselves mainly under the logic of social relations strongly inspired by the war model: the "friend/enemy" relationship. And he continues, "ethnographic fieldwork in Argentine military institutions still needs to overcome a strongly rooted social and institutional logic that assumes the existence of people who are either allies or opponents (Badaró, 2009, p. 189). Although the friends/enemies logic mentioned by Badaró has diminished, it is still present in the military, and it becomes stronger depending on the political leaning of each government administration. For example, governments that are perceived as left-wing will be more rejected resisted than others.

On the other hand, some organizations and/or social movements are a priori identified as "enemies". For example, human rights organizations and the feminist movement are perceived as being opposed to institutional values and interests. And social science professionals could be considered suspected of being leftist, and are therefore seen as less legitimate than other professionals. Anthropological research introduces, according to Badaró, another problem: "the transformation of the military into an otherness", that is, the problem of exoticizing the military, and running the risk of naturalizing the military's point of view, which holds that the army is a unique place to belong to, and too different from the rest of society (Badaró, op. cit. 193). So, how is fieldwork in the military possible in Argentina? It is possible owing to the fact that academic research can display their current activity, and weaken the dictatorial stigma of the Armed Forces? In the last decades, the Armed Forces have been interested in improving their image in society, and anthropological work could become a medium for them to save face (Badaró, 2009; Frederic et al., 2015).

What about gender in the military?

At the moment, the gender policies in the military in Argentina are the most developed in the South America, as pointed out in the introduction. The implementation of gender policies began in 2006 in a context of a new change in the civil-military relations, where the human rights agenda, which included the gender agenda, was a priority. The civilian control of Armed Forces was a crucial matter since the restoration of democracy in 1983. As Pion-Berlin states, After a prolonged period of authoritarian rule, the rebirth of democracy in Latin America has prompted a resurgent interest in civilian control of the Armed Forces. Most agreed that political leaders should resolve 'la cuestion militar' to consolidate their democratic gains (Pion-Berlin, 1991, p. 543). Ruth Diamint (2008) refers to the civilian control of the military in Argentina as "The Endless Story".[1] A new chapter of this story began in late 2005 and included a strong emphasis on gender policies, which have been part of a series of measures aimed at achieving the democratization of the Armed Forces. In this context, gender policies were part of a broader policy for the respect for human rights.

The advocacy of human rights in Argentina has the singularity of being strongly associated with the violation of human rights during the last military dictatorship. Therefore, from the military point of view, "human rights" are almost synonymous of "trials for crimes against humanity" involving the military who participated in the last military dictatorship, and as a result, they perceive human rights as a threat. It is important to know this background to understand not only the complexities of the impact of social science research on the military in Argentina, but also the historical particularities that constrain it. In this context, building bonds of trust, which have been seriously damaged throughout history, is extremely significant.

The policy of inclusion of women in areas formerly prohibited to them in the military, as well as other measures related to gender policies, was interpreted by male military members as a way to "destroy the military". Gender policies were seen, at the time, as a form of "interference" in the armed forces by the Ministry of Defense.

Similarly, the feminist movement or feminists in general are regarded, by the military, as part of the left-wing ideology, and the military′s mental representation of feminism is associated with the most radical branches, such as the feminists who make naked demonstrations to defend their rights. This characterization was radicalized in 2019, during the parliamentary debate on abortion rights, where feminist women identified with green scarves, demonstrated in the streets, in support of abortion rights. Some of the military say they agree with gender issues but not with "the feminists". Gender issues, although resisted, have a degree of acceptance that can be ascribed to the UN's gender training requirement for deployment in Peace Missions.

Method

To think about the impact of social research on gender in the Armed Forces I was inspired by the positional analysis proposed by Luc Boltanski. I will point out, among the methodological observations made by the author, those that I consider useful for my analysis. The author uses the positional analysis to measure the "power" of the ruling class to gain positions. However, I will use Boltanski's observations to analyze which institutional positions can foster the interaction between social science research and the military. That is, the possible points of contact of the interaction between two "institutional worlds" which have been historically antagonistic and only just a few years ago have engaged themselves in dialogues that were not always voluntary. At the same time, my holding of these institutional positions enabled a deeper analysis.

Another important Boltanski contribution, for the present study, is the remark about degrees as signs or emblems by which the holding of certain positions in certain fields is publicly and officially expressed. He distinguishes between institutional selection of degrees and the type of selection that each agent makes among his/her own degrees according to the expectations of the institution. For Boltanski, it is a matter of ranking the officially mentioned positions according to the degree to which the type of legitimacy they confer can be transferred to the institution and increase or intensify it by endorsing the legitimacy that it holds in its own right (Boltanski, 1973, p. 7).

According to Boltanski,

> In order to see the correlative social benefits of possessing a large social surface, it is important to remember that among the set of privileges that are the instrument and product of power, there is probably none more important than the capital of relationships. (Boltanski, 1973, p. 10)[2]

With regard to the multiplicity of positions and power, Boltanski believes that a social field can be defined as a structured whole, and, as such, endowed with specific properties of positions whose characteristics and particular properties can be analyzed, regardless of the individuals who occupy them in the way (Boltanski, 1973, p. 13).

Finally, I was interested in pointing out the relationship between the multiple institutional positions and the distance to the role. According to Boltanski, the plurality of positions that an individual holds should not make us forget that different positions are held by the same individual. Individuals are able to generate alternately, in different places and at different times, a plurality of different social personalities, whose characteristics emanate from the social position in which their holder is momentarily placed (Boltanski, 1973, p. 15). For Boltanski, there is nothing more generally, nor more functionally associated with multipositionality than the "distance to role" as a particular modality of exposure to "role" – role being defined as the set of social signum and emblems through which the legitimate occupation of a given position is publicly expressed.

To strengthen this methodological approach, we will adopt the perspective of Durão and Seabra Lopes (2011) who consider that

> looking at institutions from Hubert and Mauss's perspective (as public rules of action and thought) may help us overcome the informal/formal, human/non-human, emotional/technical and other similar dichotomies that still mark a good part of contemporary research in organizational settings, by integrating both sides in the same analysis instead of privileging either one or the other. (p. 374)

The authors consider that "this may be done by a proper focus on the environment as the locus where personhood, emotions and skills are mixed together with institutions and institutional configurations, all mutually constituting each other".

Institutional positions

As mentioned in the introduction, I aim to explore my professional performance both in the academic field and in the State bureaucracy inspired by positional analysis. The multiple positions I held constitute a privileged point of view to understand the impact of social research on the military. In agreement with the editors of this volume, by impact I mean simply having a marked effect or influence on changing military policy, practices, knowledge, skills, behavior, or living conditions (Eyal Ben-Ari et al., 2020). In Table 8.1 I list the institutional positions I have held as a scholar and in the government bureaucracy in the Defense area. For each of them, I have defined the role related to the position and the values ascribed to it and, in the last column, I point out the skills and emotions each role entails. I will briefly describe each of these columns below.

Multiple institutional positions

The first column of the table shows the different institutional positions from which I related to the Armed Forces and still do. My first approach to the military was as a university researcher. Initially, in 2007, I was invited to become a member of the Gender Policy Council for Defense of the Ministry of Defense as a gender expert on behalf of my university. This initial position was followed by the other ones. The passage through several institutional positions allowed me to gain the necessary knowledge to hold the subsequent ones. Likewise, the passage from one position to another made possible the building of a capital of social relations that Boltanski considers as probably the most important of the correlative benefits of a large social surface that an individual is capable of occupying, and that allows him or her to move through it.

Table 8.1 Institutional positions hold by the author as a scholar and in the government bureaucracy in the Defense area.

Multiple institutional positions	*Role power and/or prestige*	*Skills, emotions and personhood*
University researcher	Research on Armed Forces, gender and feminism topics	To obtain, produce and transmit scientific knowledge about gender and the Armed Forces
Advisor	Government Advisor in military training and gender	To transmit knowledge about the characteristics and identity of the Armed Forces
Lecturer	To transmit knowledge	To transmit knowledge about gender and gender in the military
Researcher at the National Defense University	Research on Gender and Military – To produce knowledge on gender in the Argentine Armed Forces	Produce and transmit scientific knowledge about gender and Armed Forces in Argentina
Gender Policy Director	Development, implementation and monitoring of gender policies Promoting institutional changes	Strategic evaluation and decision-making. Assume a place in a hierarchical structure Exercising authority Managing civil-military relations

Then, in 2008, I was hired by the Ministry of Defense, as part of a team, to carry out research about the military profession.[3] In 2009, part of this team was asked to join the staff of the Ministry of Defense. The director of the research was appointed as *Undersecretary of Training* and I was called in as an advisor. In 2010, I began lecturing on gender in the Air Force Faculty, and I have remained in that lecturing position up to now. Between 2010 and 2020, I published some articles on the incorporation and integration of women to the Armed Forces and I remained in my position as a researcher and professor in a public civilian university. In 2018, I won a grant from the Defense University for a research project on evaluating gender policies in the Armed Forces. Finally, in 2020, I was appointed Gender Policy Director at the Ministry of Defense.

Role, power, and prestige

The first point to highlight, in the analysis of the second column of the table, is the prestige and legitimacy of university degrees which, in Boltanski's words, work as signs or emblems. In addition to the degrees obtained, the position of researcher and professor in a National University adds to the construction of prestige and legitimacy. Equally important is the specialization in gender and public policy, which is legitimized through academic publications. Thus, part of the second axis of the table refers to the degree of legitimacy and/or power granted to the possession of academic knowledge and the experience in social research in gender policies, in the Armed Forces, and in both gender and Armed Forces. This background, in Boltanski's words, provides the individual with both a base of social relations and the knowledge obtained in the journey through University. Ben Ari et al. (2020) discuss part of the dynamics of academic prestige when they refer to the fact that

> the very participation of scholars and the presentation of research may act to ornament a certain meeting. In some cases, intellectuals and professors may be used simply as a background for politicians or senior civil servants to award an academic aura to it. (Ben Ari et al., 2020)

However, with the change of position and the taking on of new roles, it is required that this knowledge, acquired by the individual and legitimized by the University as an institution socially recognized for such purposes, acquires the specificities of the new professional profile. The attributes of the role will be built in the workplace itself. In this case, the passage from the advisory role, where the academic aspect prevails, to the role of public servant will require taking on a position of hierarchy and the development of competencies for decision-making and management skills. I will refer to this point later, in the section "From lecturer to policy-maker".

Skills, emotions, and personhood

The third axis of analysis refers to the capacities and emotions of individuals when they hold each of the different institutional positions. Skills, emotions, and personhood, relate to the intersubjective aspects and the possibility of establishing dialogues and meeting points between two worlds that, in Argentine history, have remained antagonistic and see each other as alien worlds.

The fact that I had carried out field work and produced knowledge about the Armed Forces (first about the military profession, and later, about gender) (Masson, 2010a, 2010b, 2017, 2019), where many military men have felt themselves reflected in ethnographic descriptions, allowed me to build a bridge between the two worlds.

When I took over as Director of Gender Policy, a woman officer who was retiring said to me: "They know you know". This sentence made reference to her military peers and meant that I would have a plus of respect due to my background of working in the field with the Armed Forces to study the military profession, my research on gender and military and my work as advisor in the Ministry of Defense (2009–2010 and 2017–2019). But it was also a guarantee for them that I knew the military world.

How does research impact?

Research with a gender perspective was started in Argentina in the 1980s at universities, institutes, and non-governmental organizations. This was one of the ways of building knowledge on gender issues (both key concepts and gender relations in the labor, family, and health fields, among others) that legitimized and gave theoretical support to the foundations of public policies with a gender perspective in the Armed Forces. This research had a positive, albeit indirect, impact on the organizational and operational dynamics of the institution and on the career development of military women. The beginning of the implementation of gender policies in the area of defense took place almost simultaneously with the development of the first piece of social research that involved direct contact, that is, ethnographic work, with the military (Badaró, 2009; Frederic et al., 2010; Frederic et al., 2015; Guber, 2016).

The previous construction of social science knowledge gave gender policies legitimacy, and allowed them to be supported by scientific knowledge and specialists in the field. The construction of socially shared meanings, promoted by an incipient feminist movement that developed after the last military dictatorship, gave them intelligibility by placing them in a local genealogy (Masson, 2015). The knowledge produced has been materialized in the production of public universities and non-governmental organizations, in the international, regional, and national norms linked to human rights and women's rights. Feminist researchers, with recognized trajectories at the local, regional, and international levels, are the people with authority on the subject. Below, I discuss four examples from which we can assess how research and researchers have impacted institutional gender transformations in the Armed Forces.

1. **Gender Institutionalism in the Ministry of Defense: Gender Policy Council and Gender Policy Directorate**

   Gender perspective in the Argentine Armed Forces began with an academic event where social science researchers showed their research work. When the first female Minister of Defense of the Republic of Argentina, Nilda Garré (2005–2010), took office, she announced the Seminar "Women and their struggles in the Argentine history" would be held in the commemoration of the International Women's Day. The result of this academic event, held at the headquarters of the office of public administration, was the publication of a book by the same name by the Ministry of Defense.

   Another landmark was the creation of the Observatory on the Integration of Women in the Armed Forces. The founding government resolution mentions the need to study the direction of women's professional careers with a view to better compliance with the national defense policy. Its functions include the diagnosis based on the production and processing of regulatory, statistical, and qualitative information on the integration of women into the military profession and the formulation of recommendations for the design of policies. The first work of the Observatory was, in 2006, a survey and a progress report. The consequence of this evaluation task was the creation, in 2007, of an Advisory Council to the Minister of Defense: The Council on Gender Policies for Defense (CGPD). The data of the Observatory are quoted in the resolution of its creation "in accordance with what results from the first report of the Observatory, a forum must be guaranteed in which women in the Armed Forces can put forth their perspectives on gender issues in the military sphere". The results were also used to define the first topics to be discussed at the meetings of this Council.

A year later, the Directorate of Gender Policy was created; it reports to the National Directorate of Human Rights and International Humanitarian Law of the Ministry of Defense. This Directorate became responsible for gender policies, coordinating its actions with the Armed Forces. The current competencies of this Directorate are to coordinate the necessary actions for the design and execution of plans, programs and actions of diffusion, sensitization and training for the incorporation of the gender perspective in the design of the public policies of the Defense Sector; to intervene in the analysis of the normative and institutional reforms in order to dictate the adequacy of the actions and institutional measures to the demands of integration of the gender perspective; to intervene in the attention and follow-up of complaints and/or petitions that involve assumptions of discrimination based on gender; to assist the National Directorate in the actions related to the inclusion of international regulations on gender in the field of international peace operations and in everything related to the adoption of measures aimed at achieving adequate protection of motherhood and paternity; and to propose protocols for the attention of cases of gender-based violence. The Directorate of Gender Policy also chairs the CGPD.[4]

2. **Grounds and rationale for institutional change: gender policy council debates and ministerial resolutions**

As mentioned above, the Argentine Ministry of Defense created the CGPD in 2007. The Council meets once a month and its function is to advise the Minister of Defense. Its creation took place in a context of multiple reforms that marked the strengthening of the civilian governance of Defense and was based on the values of modernization and democratization of the Armed Forces. Its creation resolution (274/2007) mentioned that "in view of the relevance of the issue, it is considered that a broad context, and with representation of a plurality of public actors, will promote the development of a more democratic work".

The incorporation of academic gender specialists into the CGPD shows both the interest in including expert knowledge into the debate and to add the representation of institutions such as universities. The Council was also made up of NGO representatives, who similarly contributed with knowledge and expertise built up in the non-governmental organizations spaces of the feminist movement.

Military women (officers and noncommissioned officers) also were members of the CGPD. The ministerial resolution that created the Council stated that "a space must be ensured for women in the military to put forth their viewpoints on gender issues in the military". Finally, representatives of other state agencies, such as the Ministry of Women, Gender and Diversity were also part of the CGPD. In this regard, gender policies in the Armed Forces were strongly based upon the meanings and knowledge previously built in the civilian sphere and through the interaction between the civilian and military through an institutionalized forum. Seizing this knowledge from the civilian world, some of the demands of the feminist movement were incorporated into the Armed Forces and the formulation and implementation of gender policies were legitimized.

From its very beginning, the CGPD has proposed the creation of numerous resolutions aimed at achieving equal treatment and equal opportunities for men and women in the Armed Forces. Much of the substantiation introduced into the debates came from gender research. I will briefly refer to the rationale that legitimized the decisions emanating from the CGPD.

Historian Joan Scott points to the fact that political struggle is most intense when the problems being debated cannot be justified in nature. Therefore, it is necessary to offer a solution by means of the law, which comes to replace "the truth" and to act as a guide for human action. But this substitution is not recognized as such, and the law is presented as being based on nature or on truth. That way, whoever succeeds in the political dispute manages to hide it and show his victory as a consequence of "truth" or "nature" (Scott, 1996). In the case analyzed below, we can say that the opposite took place.

The "authority reasons" to support institutional transformations were made with reference to political struggles, rights, democracy, development, and the rationale of feminist and gendered academic research. These foundations are used to reinforce the text of the norms emerging from the CGPD debates. In some cases, female academics are explicitly mentioned among the preamble to the Resolutions.

For example, the Resolution that abolishes the possibility of sanctioning people who have children out of wedlock, unmarried cohabitation, or the pregnancies of "single mothers" quotes the following paragraph from a book by sociologist Elizabeth Jelin reads: "We live in a world in which the three dimensions that make up the definition of family (sexuality, procreation and cohabitation) have undergone enormous transformations and have evolved in divergent directions. This transformation, for others called crisis, has led to a redefinition of roles within the traditional nuclear family, on the one hand, and the emergence of other forms of family organization, on the other".[5] In this case, the participation of scholars served to legitimate a certain policy or practical change.

Not only was the participation of female researchers in the CGPD sustained over time, but also the knowledge based on social science research, whether from universities or NGOs, was used in the trainings that were conducted both within the CGPD and from the Directorate of Gender Policy to members of the Armed Forces. In addition, academic specialists were called upon to participate in events, seminars and conferences and as facilitators in training workshops.

3. **Education and training of military staff**

The training with a gender perspective in the Armed Forces started with an intensive training of military women, namely the members of the CGPD. I was in charge of this training, as a gender anthropologist and a member of the CGPD. It was a one-week course, morning and afternoon, held in 2007. It included a week of work with activities such as lectures, reading, and discussion of academic texts and the submission of a final paper. In the CGPD, the focus on the training was paramount, and it was regarded as one of the key areas of work, along with the review of regulations and advocacy strategies within the Armed Forces.

The gender training was then extended to the staff of the Gender Offices and the Domestic Violence Teams through workshops. Subsequently, these same people held training talks in their own units, where the academic emphasis gradually diminished.

Later, through the exchange between academics and military women members of the CGPD, these trainings were redefined as a curricular course. In 2010, the Diploma in Gender and Institutional Management was created in the Air Force Faculty and was directed by a feminist academic lecturer and coordinated by a woman officer of the Air Force. It is a seven-month program for civilians and military personnel.

In its first year, this course had 46 students, in 2015, 81 students enrolled in it, and in 2020, the number of students enrolled rose to 150. Among those who decided to take the Diploma course, the average age is between 37 and 43 years old. For the military personnel hierarchies, these ages are placed between former Junior Officers and Chief Officers. This means that those who are interested in receiving training can be found in the middle strata of the military hierarchy. If we consider only the students belonging to the Armed Forces, the highest percentage corresponds to the personnel of the cadre of NCOs (49%), followed by the cadre of Officers (31%) and then the civilian personnel (20%) (Sotelo, 2020).

In 2020, with the arrival of the new government, as Director of Gender Policy at the Ministry of Defense, I fostered, along with the National Defense University, a special course for civilian and military staff of the Armed Forces. It is an intensive four-month university course called "Transversalization of the Gender Perspective in the Armed Forces". This course arose from the need of the Directorate of Gender Policy to provide theoretical and practical tools to the members of the interdisciplinary teams of the Gender and Domestic Violence Offices throughout the country. The objective was to guarantee a rights perspective in the implementation of gender policies and to provide skills to reverse situations of inequality, with special emphasis on cases of gender-based violence. This course is mandatory for those who implement gender policies in the Armed Forces.

In this case, academics with a gender perspective are once again more strongly present in the Armed Forces. A very relevant fact is that out of the four lecturers belonging to the Diploma course, three of them are civilian and military personnel of the Armed Forces specialized in gender and only one does not belong to the Forces. Table 8.2 shows the number of people who have enrolled for the Diploma course, disaggregated by Force and by sex.

Table 8.3 shows the number of people enrolled, disaggregated by sex and by officers, non-commissioned officers, civilian staff and volunteer troops.

Table 8.2 People enrolled for the Diploma course, disaggregated by Force and by sex.

Enrolled	*Army*	*Navy*	*Air Force*	*Total*
Women	106	13	23	142
Men	48	7	2	57

Table 8.3 People enrolled, disaggregated by sex and by military or civilian status.

Enrolled	*Officers*	*Non-commissioned Officer*	*Civilian personnel of the Armed Forces*	*Volunteer soldier (paid)*	*Others*
Women	51	22	58	5	6
Men	37	7	12	1	0
Total	88	29	70	6	6

This Diploma course is the result of political will and new commitment to the insertion of academic knowledge with a gender perspective in the Armed Forces. The novelty of this second step in gender policies is that the Armed Forces already have among their ranks highly trained women who may themselves become teachers of the Diploma course.

4 The construction of reflexive knowledge: academic experts and the military

A new step forward in the influence of social science research with a gender perspective in the military was the submission of research projects at the Universidad de la Defensa Nacional. In 2017, two projects were funded: Evaluation of Gender Policies for Defense: Advances, Obstacles and Challenges (2007–2017) and Prejudices, Gender Stereotypes and Associated Beliefs: An Analysis with Students of Military Education. The first one included academics and military women and its objective was to analyze and evaluate the impact of gender policies in the Armed Forces since the creation of the CGPD until the year 2017. Just as the origin of the first Diploma course arose from the confluence of the academic and military worlds in the CGPD, this research project arised in the context of the Diploma course, also from the joint work of feminist academics and military women.

The result of this research work was published by the Ministry of Defense in 2020 under the title *Militares Argentinas*: *Evaluation of Gender Policies in the Defense Area*. The book is divided into three parts: Gender Policies in the Defense field, Institutional instruments to address situations of violence against women, and Experiences of integration of women into the Armed Forces. Many different authors contributed to its chapters: academic members of the CGPD, advisers of the Directorate of Gender Policy, military members of the Army, Navy and Air Force and students of the Diploma in Gender and Institutional Management. I directed the research and compiled the book. All this took place before taking up my position as Director of Gender Policy.

This project shows that the interaction between female scholars and military women took the shape of a virtuous spiral in which the initial institutional mechanisms (the Observatory of Women in the Armed Forces, the CGPD and the Directorate of Gender Policy) generated opportunities such as the Diploma in Gender and Institutional Management, which in turn encouraged social research in the Armed Forces.[6] In this case, the level improvement of in-house research by setting academic standards is seen as an indirect effect. The writing of some of the chapters by the military and the engagement in the research involved a reflective work that brought them closer to the logic and concepts of the social sciences.

As a piece of research that provides a richer understanding of the Armed Forces and promotes the application of theoretical gender frameworks to the institutional dynamics of the Armed Forces, the process of researching and writing the book can be considered an indirect influence on various actors. This impact is related to the enlightenment and critical sociological models that make possible a raised awareness of social issues emerging out of investigations and which are politically important for activists such as human rights lawyers or feminists (Ben Ari et al., 2020). These research findings are also being used today for the training of military staff.[7]

From lecturer to policy-maker: challenges of the new position

I will now analyze the possibilities of impact of social research when a lecturer occupies a decision-making position such as a hierarchical position within the State bureaucracy. To this end, I will reflect about my experience as Director of Gender Policy at the Ministry of Defense. I will also return to Boltanski's conceptualization of "role distance" which the author considers as a necessary counterpart to multi-positioning and the permissive condition for the harmonious handling of the different social personalities which the individual brings together, particularly by making the passage from one to the other somewhat painless (Boltanski, 1973, p. 17). As mentioned above, Boltanski emphasizes that individuals possess the ability to perform in a plurality of different ways. They are able to generate alternately, in different places and at different times, a plurality of different social personalities, whose characteristics derive from the social position they are momentarily placed in. According to the author, if we thus agree to define the individual as the convergence of all the socially required personalities that he or she is able to generate, we can see that the management of an "individual group" requires the implementation of complex strategies necessary to produce the right personality for each place and time or to prevent maladjusted personalities from meeting and dialoguing (Boltanski, 1973, p. 16). In other words, the change of position requires the ability to vary manners, style, discourse, skill, emotions, behaviors, and all social performances according to the environment, the place, or the group.

The position of Director of Gender Policy calls for skills from the political field which are very different from those required by the academic field. Following Boltanski's analysis, university members, even when they hold several positions, remain mostly confined to the academic field, and, at best, to the intellectual field. In his analysis of the multiplicity of positions of power, the author also shows that

> The more autonomous a field is, the less the holders of local power have (as is the case in the academic or intellectual field) an extended social surface and general power; conversely, the less autonomous a field is, the more the occupation of positions of power in that field includes the occupation of positions of power in other fields, the increase in local power being correlative here (as is the case in the political field). (Boltanski, 1973, p. 14)[8]

The challenge then is a) to transfer the strengths – those which permit it – from the academic to the political field; b) to transform those that can be transformed in order to respond to the requirements of the new position; c) to acquire new skills that are characteristic of the policy-maker position and d) to expand the social relations capital, which is one of the ways to achieve greater power (in the same way that the previously acquired power also depends on the relations capital). In that sense, the passage from a position in the academic field to a position in the political field will highlight the differences between the two and the weaknesses that will need to be strengthened. When a lecturer takes up a position in the State bureaucracy – as in my case in the Ministry of Defense – he or she leaves "the position situated in these tabernacles of 'disinterestedness' and 'public service', in short, of the neutrality and propriety that is the university field" (Boltanski, 1973, p. 8).

When the individual moves from the position of researcher, lecturer and/or advisor to the position of authority (policy-maker), there is a change in the position of power. The great challenge of multi-positionality, in the case of the researcher-lecturer-advisor versus the decision-maker, is the transformation of the nature and the source of its power. In the first case, power is associated with the prestige of erudite knowledge, which other institutional actors do not possess and rely upon (and therefore the researcher is admired for this socially valued knowledge). Yet when he or she occupies a position of policy-maker, the power is associated with the functions of the position he or she holds. In this case, the principle of hierarchy and authority inherent at State institutions, and more specifically in the case of the Armed Forces, takes effect. The scholar must become able to exercise power, understood as the capacity and management of strategic resources to make other institutional actors act in a certain way.

What are the challenges when a new role or social personality must be performed in the same institution where the lecturer had developed bonds of trust as an academic expert and alien to bureaucratic logic? What positive aspects of the previous position can be transferred to the new position and which ones may be affected by the characteristics resulting from the new role?

Tensions, weaknesses, and opportunities in the new position

Implicit in my position as an academic was the listening, knowledge and objective understanding that led to certain institutional activities where a few women in the military were involved. This allowed me to gain a deeper understanding of certain aspects of the institution, expand the network of relationships and strengthen my academic position in the defense field through teaching. As for them, my knowledge allowed them to support their career advancement, providing solid grounds, support for their concerns or proposals, and sometimes it was simply my presence as an academician that legitimized certain spaces or events. When I took up the position of Director of Gender Policy, it was necessary to play a new role, understanding role as the set of social signum and emblems through which the legitimate occupation of a given position is publicly expressed. I will analyze the opportunities and weaknesses of the new position through the four challenges mentioned in the previous section:

a. transferring the strengths of the academic field to the political field

   One of the characteristics that continued to be valued by the military was the prestige of knowledge. As I mentioned earlier, this situation can be summarized in a phrase that I heard repeatedly when I took up the position of Director of Gender Policy: "They know you know". In some way, the knowledge accumulated through social research on the military diminished the distrust of the intervention from a position of "outsider". This contributes, and in fact did, to the possibility of establishing a dialogue, assessing changes in collaboration and implementing some institutional changes in a coordinated way.
b. transforming the strengths that can be transformed to meet the requirements of the new position

   One opportunity is to transform academic knowledge about institutions into knowledge applicable to public management in order to anticipate the possible resistances and impacts that the application of public policies may generate. In other words, to understand the institution in order to know what decisions must be made, which actors to engage in dialogue with, and what the field of negotiation is, according to the institutional times within which public policies are decided.
c. to acquire new skills appropriate to the position of policy-maker, understanding the exercise of power within a hierarchical bureaucratic institution

This is the process in which I have experienced the greatest tensions and resistance, as it required the acquisition of new skills, ways, styles, and discourses toward which some institutional actors with whom I had worked from a lecturer position have been refractory. This was especially true in those cases where it became evident that the role of lecturer or advisor had given way to the new policy-making function which required on their part a response to the logic of hierarchy and authority.

d) to expand the social relations capital as one of the ways to achieve greater power

The new institutional position opens up possibilities for the expansion of the "social surface" through new social relations. In this vein, the capital of relations is linked to the characteristic of the new position that includes, within the skills of the policy-maker, a field of relations proper to the position he or she occupies. In other words, the position itself prefigures a field of relations according to that which defines it in terms of bureaucratic structure.

Conclusions

By way of conclusion, I will refer to some of the three orientations toward knowledge related to the way in which social research impacts on the Armed Forces, mentioned by the book's editors (explicit and implicit objectives, direct and indirect influence, and intentional and unintentional consequences). Finally, I will add a brief reflection on the intersection of academic knowledge with the requirements of a policy-maker's position.

I will first refer to the explicit objectives and the indirect influence. Here we can highlight the impact of social research on gender (not related to gender and the Armed Forces) in the first stage of the implementation of gender policies in the Argentine Armed Forces. At that time, there was an effort and an explicit objective to transform the internal norms to adapt them to national and international standards and to guarantee equity between men and women in the Armed Forces. Social science research was useful to legitimize the changes in the official documents, but also to argue about them in the debates of the CGPD. An articulation between explicit objectives and indirect influence is the research carried out as a "situation diagnosis", whose results are then cited as the basis for the implementation of institutional gender areas, as is the case with the creation of the CGPD.

Among the indirect influences is also the position of professor of the Gender Degrees, where the classes and the reading of some academic works have modified the students' perspective on gender relations and public policies with a gender perspective. Some young women officers have radically changed their outlook based on concepts that they found enlightening, for example, the fact that institutions are not gender-neutral. Another instructive experience was the invitation, as an expert, to deliver talks to a unit where a person had effected the change of her gender identity. The Head of Personnel had been a student in the Gender Diploma program and said that thanks to my classes, he had understood what gender was all about.

I have also observed the use of analytical concepts from recent research, for example, between recruitment and integration of women into the Armed Forces. This distinction was made by a General at the presentation of the book *Militares Argentinas* at the Army NCO School. These are formulations that allow officers to better understand why gender is still "talked about" and, perhaps most importantly, to internalize the subject and embrace it. This was made possible by the participation of a female military officer as one of the authors of the book. The shared construction of knowledge between academicians and military allowed, in this case, the appropriation of that knowledge as belonging to the Armed Forces. It is no longer the feminist academics speaking, but rather the military women.

In terms of informal objectives, gender policies have been both an opportunity and a potential disadvantage for the military and in some ways also for academicians. For women in the military, there has been an opportunity because they have allowed them to occupy previously unimaginable institutional positions. They became a liability when political support declined and there was some "revenge" from some commanders. For academicians, it means producing knowledge in a field that is little known as a positive aspect and at the same time, as I mentioned above, the risk of "contamination" or the accusation of being "co-opted" by the military. The informal objectives became visible in my experience as I moved from a position of ally as an academic expert to a position of power and hierarchy. This has produced stress points where previously there were implicit alliances.

Finally, I would like to reflect on the section "From scholar to policy-maker: challenges of the new position", which deals with an institutional position where academic training and the functions of a policy-maker converge. I believe that the move from one position to another poses challenges that include both some weaknesses and opportunities. While some strengths of academic knowledge, such as legitimacy and prestige, can be transferred, some skills will need to be transformed and others learned. Success in resolving these challenges will determine the degree to which the academic knowledge acquired can have an impact on the military.

Notes

1 This brief summary gives an overview of the expression used by Ruth Diamint "Argentina is the Latin American country that has made the most progress in civilian control of the Armed Forces. It is the first country to bring to justice those responsible for the dictatorship and one of the few where military intervention in domestic security is clearly prohibited. The design created in the first years of the transition to democracy aimed at juridifying the civil-military relationship, so as to clearly separate the right from the illegal and mark a break with the coup tradition. This path, however, was marked by advances and setbacks, military insurrections and the search, by the Armed Forces, for new spaces of autonomy. Although the government of Néstor Kirchner re-established a human rights agenda and implemented some important measures, the institutionalization of the new role of the military is far from having been consolidated" (Diamint, 2008, p. 95).

2 Boltanski defines "social surface" as the portion of the social space that an individual is able to traverse and control by successively occupying (albeit often at a very rapid pace) the different social positions that he would be able to occupy simultaneously on the sole condition of physically possessing the gift of ubiquity that is socially conferred upon him (Boltanski, 1973, p. 11).

3 The first investigation was at the request of the Ministry of Defense. Every time we carried out field work, two people from the research team would travel to the units: a woman and a man and, in some cases, two women. At that time, the resistance to our presence was linked to the fact that the work was commissioned by the Ministry of Defense, which, reproducing the civil-military dichotomy, is seen by the military as an institution outside the Armed Forces and therefore

unable to understand military logic. Once the research had been completed and the report presented, that view changed because many members of the military saw themselves reflected in the results of the research, where their opinions and voices were expressed. This gave us credibility and confidence in their eyes.

4 The decision about who occupies the position of Director is usually agreed upon with representatives of the feminist movement who also belong to the governing political party. So far feminists have been referents of academic feminism and NGO feminism. The Directorate of Gender Policies is a space that is relevant both to political parties and to the feminist movement.

5 Quote from *Pan y afectos. La transformación de las familias* published in 2000 by Fondo de Cultura Económica.

6 This piece of research was carried out by members of all the institutional mechanisms previously mentioned, except for the Observatory, which set the events in motion.

7 The book *Militares Argentinas*, is part of the reading requirements of the Diploma Course "Transversalization of the Gender Perspective in the Armed Forces", which is compulsory for the Gender Offices and THE Domestic Violence Teams.

8 Boltanski suggests that the social surface of an individual depends as much on the dispersion of positions between different social fields as on their number. If we take into account that, according to the author, the academic field is one of the most autonomous and that lecturers tend to occupy positions within the same field, the social surface of lecturers tends to be reduced. In order to have an extended social surface, it is necessary to occupy positions distributed in different fields. The correlative social benefits of possessing a large social surface, as the author suggests, result from the fact that among the set of privileges that are an instrument and product of power, there is probably none more important than the capital of relationships. In other words, the greater the "social surface", the greater the capital of relations and the greater the possibilities of building and exercising power.

References

Badaró, M. (2009). *Dilemas da antropologia das instituições controvertidas: reflexões a partir de uma investigação no Exército argentino*. In C. Castro, & P. Leirner (Eds), *Antropologia dos Militares. Reflexões Sobre Pesquisas de Campo.* [Anthropology of the military. Reflections on field research]. Rio de Janeiro: FGV Editora.

Ben Ari, E., Carreiras, H., & Castro, C. (Eds). 2021. *The Impact of Social Research on the Military: Reflections and Critiques.*

Boltanski, L. (1973). L'espace Positionnel: Multiplicité des Positions Institutionnelles et Habitus de Classe [Positional Space: Multiplicity of Institutional Positions and Class Habits]. *Revue française de sociologie* 14 (1): 3–26.

Diamint, R. (2008). La Historia Sin Fin: El Control Civil De Los Militares en Argentina [The Never-ending Story: Civilian Control of the Military in Argentina]. *Revista Nueva Sociedad*, No 213, enero-febrero. 95–111.

Durão, S., & Seabra Lopes, D. (2011). Introduction: Institutions Are Us? *Social Anthropologie/Anthropologie Sociale*, 19. European Association of Social Anthropologist.

Frederic, S., Graciano, O., & Soprano, G. (2010). *El Estado Argentino y las Profesiones Liberales, Académicas y Armadas* [The Argentine State and the Liberal, Academic and Armed Professions]. Prohistoria ediciones.

Frederic, S., Masson, L., & Soprano, G. (2015). *Fuerzas Armadas en Democracia. Percepciones de los Militares Argentinos Sobre su Reconocimiento* [The Armed Forces in Democracy. Perceptions of the Argentine Military Regarding Their Recognition]. Prohistoria ediciones.

Guber, R. (2016). *Experiencia de Halcón. Ni Héroes ni Kamikazes, Pilotos de A4B* [Falcon Experience. Neither Heroes Nor Kamikazes, A4B Pilots]. Buenos Aires: Sudamericana.

Jelin, E. (2000). *Pan y Afectos. La Transformación de las Familias* [Bread and Affection. The Transformation of Families]. Buenos Aires: Fondo de Cultura Económica.

Masson, L. (2010a). Las Mujeres en las Fuerzas Armadas. ¿Qué tipo de integración?. In *Género y Fuerzas Armadas: Algunos Análisis Teóricos y Prácticos* [Gender and the Armed Forces: Some Theoretical and Practical Analyses]. Buenos Aires: Ministerio de Defensa y Fundación Friederich Ebert.

Masson, L. (2010b). La Transformación del rol de las mujeres en las Fuerzas Armadas Argentinas: Hacia la Construcción de un Espacio Mixto [The Transformation Of The Role Of Women in the Argentine Armed Forces: Towards The Construction Of A Mixed Space]. *Center for Hemispheric Defense Studies Security and Defense Studies Review. An Interdisciplinary Journal*, all-Winter Issue, 11.

Masson, L. (2015). Saberes Académicos, Experiencias y Militancias. Buenas Prácticas en Políticas Públicas con Perspectiva de Género. In *Equidad y Defensa: una política en marcha IX* [Equity and Defence: A Policy on the Move IX]. Buenos Aires: Ministerio de Defensa.

Masson, L. (2017). Women in the Military in Argentina: Nationalism, Gender and Ethnicity. In *Gender Panic, Gender Policy. Advances in Gender Research*, Vol 24 (pp. 23–43).

Masson, L. (2019). Tensions and Negotiations Surrounding The Inclusion Of Female Officers in the Brazilian Air Force. In R. Moelker, M. Andrés, & N. Rones, *The Politics of Military Families. State, Work Organizations, and the Rise of the Negotiation Household.* Routledge.

Masson, L. (Ed.) (2020). *Militares Argentinas. Evaluación de Políticas de Género en el ámbito de la Defensa.* [Argentinean Military Women. Evaluation of Gender Policies in the Defense Field]. Buenos Aires: Ministerio de Defensa, UNDEF Libros.

Pion-Berlin, D. (1991). Between Confrontation and Accommodation: Military and Government Policy in Democratic Argentina. *Journal of Latin American Studies* 23 (3), October 1991: 543–571.

Scott, J. (1996) *Only Paradoxes to Offer. French Feminists and the Rights of Man.* Cambridge, Massachusetts; London, England: Harvard University Press.

Sotelo, E. (2020). Formación en Género desde las Fuerzas Armadas. La Diplomatura de Género y Gestión Institucional de la Fuerza Aérea Argentina". In: L. Masson (Ed.), *Militares Argentinas. Evaluación de políticas de Género en el ámbito de la Defensa.* [Argentinean Military Women. Evaluation of Gender Policies in the Defense Field]. Ministerio de Defensa, UNDEF Libros.

Index

For Product Safety Concerns and Information please contact our EU
representative GPSR@taylorandfrancis.com
Taylor & Francis Verlag GmbH, Kaufingerstraße 24, 80331 München, Germany

www.ingramcontent.com/pod-product-compliance
Lightning Source LLC
LaVergne TN
LVHW021131300926
849160LV00010B/961
9780367760397